TEN ESSENTIAL SKILLS FOR ELECTRICAL ENGINEERS

TEN ESSENTIAL SKILLS FOR ELECTRICAL ENGINEERS

Barry L. Dorr

Published by John Wiley & Sons, Inc., Hoboken, New Jersey.
Published simultaneously in Canada.

Library of Congress Cataloging-in-Publication Data:

Dorr, Barry L., 1958-
Ten essential skills for electrical engineers / Barry L. Dorr.
pages cm
Includes bibliographical references and index.
ISBN 978-1-118-52742-9 (pbk.)
1. Electrical engineering–Problems, exercises, etc. 2. Electrical engineering–Vocational guidance. I. Title.
TK169.D67 2014
621.38–dc23

2013026573

10 9 8 7 6 5 4 3 2 1

This book is dedicated to my father, Roger Dorr, who shared with me his love of all things mechanical, electrical, and musical.

It is dedicated to my wife, Judy, who brings joy to me and all whom she touches.

CONTENTS

PREFACE

This book and the companion website* will do two things for you: First, it will help you excel in your technical interviews so you can secure a great first job. Second, it will help you succeed in that job by showing you how to solve real-world problems using what you already know from your engineering courses.

There is often a disconnect between what you learned in school and the expectations of those interviewing you. On one hand, the engineering curriculum has taught you a tremendous amount of theoretical and practical material, but it is simply not possible to remember it all in an interview. On the other hand, the interviewing team will likely consist of your potential supervisor and coworkers, and they will be looking for someone with the skills to be an immediate contributor in their department. The interview will typically focus on *what you can do* instead of *how much you know*. This book bridges the gap by reviewing how to apply what you learned in school to practical engineering tasks.

Each chapter reviews design skills by relating them to fundamental concepts. Consider the skill of designing an averaging FIR digital filter. This is a valuable skill, but the best designers understand the basics of digital filters—they understand the practical use of z-transforms and how to work with any discrete-time structure. As a result, they can work with a wide variety of digital filters thereby enabling them to design the best filter for a given requirement. This text reviews digital filter design and many more commonly required design skills covering a wide variety of areas and relates them to the fundamentals. Chapters begin with the fundamentals and then present the skills.

As a workplace reference, this book will give you information you need to succeed at basic design tasks. For more in-depth problems, it will help direct you to manufacturer's literature or the appropriate sections of your college textbooks. I hope it will merit valuable space on your bookshelf for many years.

*www.blog.dorrengineering.com

Whether you are a recent graduate or an experienced engineer, your time is valuable. This book was written to help you—not challenge you. To this end, the mathematical derivations are kept simple, but not simplistic, and the examples presented show practical applications using practical skills that will help you in job interviews and during your first several years in the workplace. Each chapter in this book can be read in a matter of hours—not days.

Finally, I hope that you enjoy working through the examples and problems in this book, and I hope they motivate you to tackle and solve hundreds or thousands of problems over what should be a satisfying and rewarding career.

Barry L. Dorr, PE

Carlsbad, CA
May 2013

ACKNOWLEDGMENTS

Shortly after obtaining my BSEE and moving to San Diego, I attended several signal processing lectures by Professor fred harris (capitalization omitted at his insistence) at San Diego State University. Professor harris had an uncanny ability to explain difficult signal processing techniques using basic concepts and simple drawings. Wanting to learn more, I pursued an MSEE at San Diego State and was fortunate to have him as part of my thesis committee. Professor harris has been teaching and inspiring students and working engineers in San Diego for over 45 years. I'm sure I speak for all my colleagues in San Diego when I thank him.

I was fortunate to have David Hull as a mentor at my first job. David guided me through many of my first design tasks with skill and patience. I would frequently take his lab books home to study at night and return in the morning with a list of questions which he would cheerfully answer despite his busy schedule.

I would also like to thank James Crawford who has been a mentor, a source of technical inspiration, and a valued friend for many years.

I would like to thank all of my current colleagues, especially my good friends Jeff Babb, John Huebner, Steven Chuwang, Leon Huynh, and Kim Do, who all provided excellent feedback and personal encouragement throughout the project. Working with this great group always left me with plenty of energy and enthusiasm to write late into the night. I would like to thank John Wiley and Sons for sharing my vision for this book and Ms Kari Capone for working as my liaison with Wiley. I would also like to thank Professor Dennis Derickson at Cal Poly San Luis Obispo for providing valuable feedback at the beginning of the project.

Good reviewers find mistakes and make suggestions. Excellent reviewers take great effort to understand the author's purpose and vision, and then help the author succeed with that vision. I was fortunate to have the help of three of the best.

Any interaction with my nephew, Joshua Dorr, is always a pleasure, so I was thrilled to have him as a reviewer. He provided excellent feedback on accuracy of concepts, mathematical derivations, and correct use of the

English language. I'll always appreciate that Josh agreed to undertake this project during the completion of his PhD thesis.

Bruno Paillard is living proof that there are wonderful and special people at the other end of the Internet—he lives over 3000 miles away, and we have met in person only once. But our common interest in things technical has turned into a special friendship over the years. Bruno is a practical, hands-on engineer with a comprehensive theoretical background. In addition to his technical review, his insightful comments on what was and was not relevant to engineers were invaluable in determining the scope of the book.

Jon Bernardi's background is chemical engineering, but after reading several of his excellent magazine articles, I asked him to be a reviewer. Jon's teaching and travel schedule keep him extremely busy, so he reviewed the chapters and wrote his feedback mostly from airline seats and hotel rooms. He exemplifies the saying "If you want something done, give it to a busy person." Jon masterfully removed the tension from many poorly worded sentences and also provided expert guidance on consistency throughout the project.

The reviewers are special people and special friends. Their fingerprints are all over the pages of this book.

B. L. D.

ABOUT THE AUTHOR

Barry L. Dorr, PE has designed signal processing algorithms and circuitry in San Diego for nearly 30 years. He holds a BSEE from California Polytechnic State University in San Luis Obispo and an MSEE from San Diego State University.

Mr Dorr's interest in electronics began when, as a high-school freshman, he purchased a worn-out 1949 Ford pickup truck from a Southern California orange grower. As the boy disassembled and rebuilt the truck, his father, an engineer, saw an opportunity to improve his son's less-than-satisfactory math grades by artfully explaining how each system worked, how it could be described mathematically, and how it might be improved. The old truck became a rolling laboratory for numerous high-school and college projects involving electronics, heat transfer, engine control, and troubleshooting. Working on the truck left him with a lifelong fascination with all things electrical, mechanical, and mathematical.

Mr. Dorr has worked at several small and large electronic firms in San Diego and owned and operated a small consulting firm for 11 years. He currently designs DSP and RF systems for Datron World Communications. He holds eight patents for various communication and signal processing techniques. Outside of work, he enjoys bicycling, working on cars and motorcycles, and playing trombone as a freelance studio musician.

ABOUT THE REVIEWERS

Jon Bernardi

Jon Bernardi (BS Chem Eng., University of Illinois; MBA, Bellarmine University) spent 36 years with the BFGoodrich/Lubrizol organization in a variety of roles: process engineering, project engineering, quality systems, information systems, process improvement, Six Sigma black belt improvement projects, and safety systems. This was an even split between manufacturing level and corporate responsibilities. He would be given a job and told "Work your way out of this, and we will find something else for you to do." It really worked out that way and he enjoyed a great ride. In his last corporate role he served as the Lubrizol process safety manager. He had the opportunity to work with and travel to the various company sites around the world. In conjunction with his regular job duties, he was also the in-house trainer for Cause Mapping® (ThinkReliability), a root cause analysis methodology. Now, semi-retired, he works as an instructor for ThinkReliability, teaching a wide range of companies root cause analysis for incident/accident investigation. When not in the air, he is on the road playing with his old car, seeking out new architectural sites or just relaxing with a good book.

Joshua Dorr

Joshua Dorr is a PhD candidate in physics at Harvard University. Though a physicist by trade, his interest in circuits has grown from the deep connection between electronics and measuring fundamental properties of the real world. He is currently working on measuring single cyclotron excitations of a single trapped positron at 100 mK—a project that has led him to use many of the techniques covered throughout this book. When he's not designing and building scientific apparatus, he enjoys hiking, climbing, bird watching, or any other activity that combines learning, adventure, and the outdoors.

Bruno Paillard

Bruno Paillard initially trained as a technician at IUT-Le Creusot (France), then as an engineer at INSA Lyon (France), he then went on to complete his PhD in digital signal processing from Université de Sherbrooke (Canada). His dissertation focused on the fields of auditory modeling and perceptual coding of music. From 1993 to 2003 he was a professor of electrical and computer engineering at Université de Sherbrooke, where he taught classes in embedded systems and did advanced research on adaptive filtering and active control of acoustical and vibrational processes. His research work led to several groundbreaking applications and technology transfers, including the world's first multi-channel active noise cancellation system in an industrial exhaust stack. In 1998, while still working at Université de Sherbrooke, he began consulting for D-BOX Technologies (Canada), who were developing the world's first motion simulation system for home theater applications. He was Director of Research and Development for D-BOX from 2004 to 2008 and while there developed their unique vibro-kinetic technologies. In particular, he was responsible for the hardware and software designs of the DSP systems, including breakthrough designs for the proprietary electro-mechanical motion actuators and computer-based motion controllers. His work at D-BOX contributed greatly to them being awarded over half a dozen patents for motion simulation systems and technologies. In 2008, he co-founded Convergence Instruments where he is senior designer and is responsible for all the leading-edge product designs.

NOTE TO INSTRUCTORS

Recent graduates have always been valued for their enthusiasm and creativity, and in better economic times were often seen as the future of the company. As a result, they were given some time to acclimate to their new job before being expected to do critical engineering tasks. But the combination of today's result-oriented workplace and the fact that most recent graduates will switch jobs after a few years results in companies expecting new graduates to enter the workplace with demonstrable design skills.

This presents a dilemma for academia. It must give the students the breadth, depth, and critical thinking skills they need to be engineers, but in order to get hired after graduation they will often have to compete with the practical design skills of experienced engineers. Most engineering curriculums include practical design skills, but in nearly 30 years of interviewing students, the author has found that they frequently can't demonstrate those skills in an interview.

This book addresses this dilemma. It consists of nine chapters of purely technical review followed by a chapter on interviewing strategies and thriving in the workplace. When the students read the chapters and work the problems they will get an infusion of practical skills and a strong reminder that the fundamentals of the EE curriculum can be used to solve nearly any problem they'll encounter in the workplace. The text was written to be easily readable and the problems are straightforward, so the material will help the students rather than challenge them.

This book can be integrated into the EE curriculum starting in the sophomore year. As the students approach the end of individual courses, the relevant chapters can be assigned and they'll find the material quite easy. As they progress through the curriculum, they will complete the entire book. As graduation approaches they will recall that these chapters provide a practical review and revisit them as they prepare for their technical interviews. As they prepare for interviews they are encouraged to visit the companion website www.blog.dorrengineering.com where they will find a large collection of interview practice problems and additional practical skills.

The book can also be used as a supplement to a one-quarter or semester senior seminar course. The material is not overly difficult, so the instructor can use class time to focus on other aspects of the course and have the students review this material as self or group study. The schedule outlined below will require about 4–6 hours per week of the student's time.

ONE SEMESTER COURSE

Week 1: How to Design Resistive Circuits
Week 2: How to Prevent a Power Transistor from Overheating
Week 3: How to Analyze a Circuit
Week 4: How to Use Statistics to Insure a Manufacturable Design
Weeks 5, 6: How to Design a Feedback Control System
Week 7: How to Work with Op-Amp Circuits
Weeks 8, 9: How to Design Analog Filters
Weeks 10–12: How to Design Digital Filters
Weeks 13, 14: How to Work with RF Signals
Week 15: Getting a Job—Keeping a Job—Enjoying Your Work

ONE-QUARTER COURSE

Week 1: How to Design Resistive Circuits
Week 2: How to Prevent a Power Transistor from Overheating
Week 3: How to Analyze a Circuit
Week 4: How to Use Statistics to Insure a Manufacturable Design
Week 5: How to Work with Op-Amp Circuits
Weeks 6–9: Let the students choose between remaining chapters based on individual interest
Week 10: Getting a Job—Keeping a Job—Enjoying Your Work

Students will need MATLAB® for many of the examples in this book. If it is not available through the university they can contact MathWorks to obtain a student edition or a trial edition.

If you or your students have comments or questions about the book or if you find errors please send e-mail to bdorr@dorrengineering.com.

Finally, graduation time can be a stressful time for students. In addition to completing their course requirements, they are facing many other changes in their lives. It would have been impossible to write this book without sharing my own enjoyment of being an electrical engineer. Working through these examples and problems will remind graduating seniors that they have a fascinating, exciting, and enjoyable career awaiting them.

1

HOW TO DESIGN RESISTIVE CIRCUITS

Chances are good that every schematic diagram you've seen contained at least a few resistors. This component is an electrical workhorse, commonly used for establishing bias voltages, programming gain, summing signals, attenuating signals, and numerous other functions. Ideal resistors dutifully follow Ohm's law, which has no frequency dependence, so it is easy to believe that designing with resistors is a simple task. This is probably the most common reason candidates are caught off guard and fail when asked to design simple resistive networks in interviews. This chapter will show you how to design and analyze practical resistive networks that solve problems you'll encounter in interviews and in the workplace.

The resistor was probably the first component you studied in school. At that time, it was the only component in your toolset, so the problems you solved were limited to finding voltages and currents in DC networks. Doing these problems taught you valuable skills such as nodal and mesh analysis, but the problems were not particularly practical and perhaps not very interesting. As a graduating engineer your knowledge of circuit elements has broadened

Ten Essential Skills for Electrical Engineers, First Edition. Barry L. Dorr.

significantly and you have better computer tools to help with the mathematical manipulations. The examples and problems in this chapter should be much more interesting because they represent practical design problems; they should be more enjoyable because, after setting up the problems, we will rely on the computer for the brunt of the manipulations.

This chapter begins with the commonly asked interview problem of creating a voltage source with a specified Thevenin resistance. Since this is such a common problem, we derive equations so that you can easily compute the resistor values when you encounter it. Next, we design a coupling circuit with specific design requirements. Many experienced engineers design this circuit using an op-amp and numerous resistors, but you'll see that a network with only three resistors fulfills the design requirement. We then design a 50 Ω bidirectional attenuator that is commonly found in RF circuits. This is not an easy problem, but it provides a good example of converting design requirements into solvable equations and then calculating the components. Since the attenuator design is a difficult problem, we check our result using mesh analysis, and you'll see that analyzing resistive networks is simply a matter of writing the correct equations and then letting a matrix solver compute the solution.

1.1 DESIGN OF A RESISTIVE THEVENIN SOURCE

Designing circuits involves taking inventory of what is required, what is known, what is unknown, and what is available. When designing resistive networks we must often solve simultaneous nonlinear equations as shown in the following example:

Example 1.1. Given a 3.3 V power supply, design a resistive circuit that provides 1.8 V with a source resistance of 1.5 kΩ.

Solution. This is a practical problem encountered frequently when designing bias circuits. We immediately note the desired output voltage is less than the supply voltage so a resistive circuit will suffice. We also recognize that the requirement leads us to a Thevenin source.

Knowing that two design parameters (Thevenin voltage and resistance) must be simultaneously satisfied, it makes sense[1] to try the circuit with two resistors as shown in Figure 1.1b.

The design procedure is to compute R_1 and R_2 so that the Thevenin equivalent of the proposed network in Figure 1.1b is the Thevenin source of

[1]If it *doesn't* make sense don't worry. You'll pick this up with time.

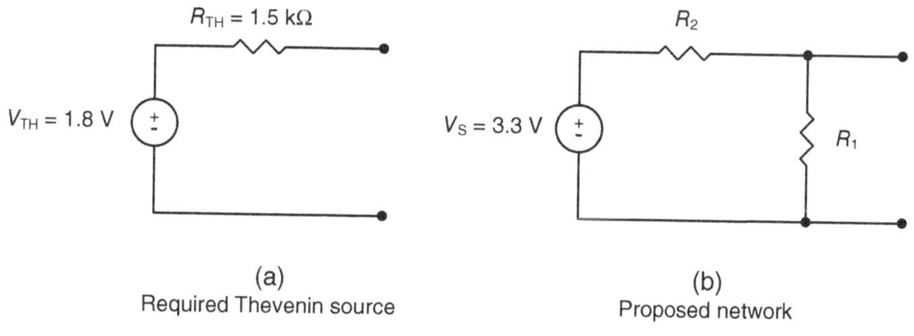

Figure 1.1. Required Thevenin source (a) and proposed resistive network (b) for Example 1.1.

Figure 1.1a. The Thevenin resistance of the proposed network is the resistance seen at the terminals with the voltage source shorted, so our first equation is the formula for parallel resistances:

$$R_{TH} = \frac{R_1 R_2}{R_1 + R_2} \tag{1.1}$$

The Thevenin voltage is the voltage at the terminals of the circuit of Figure 1.1b when it is unloaded, so we use voltage division to write the second equation:

$$\frac{V_{TH}}{V_S} = \frac{R_1}{R_1 + R_2} \tag{1.2}$$

These two equations are nonlinear and cannot be simultaneously solved using a simple matrix. Instead, we solve Equation 1.1 for R_1 which gives

$$R_1 = \frac{R_{TH} R_2}{(R_2 - R_{TH})} \tag{1.3}$$

Then, substitute R_1 from Equation 1.3 into Equation 1.2 and solve for R_2:

$$R_2 = \frac{R_{TH} V_S}{V_{TH}} = \frac{1500 \times 3.3}{1.8} = 2750\ \Omega \tag{1.4}$$

Finally substitute Equation 1.4 into Equation 1.3 to compute R_1:

$$R_1 = \frac{R_{TH}}{\left(1 - \frac{V_{TH}}{V_S}\right)} = \frac{1500}{\left(1 - \frac{1.8}{3.3}\right)} = 3300\ \Omega \tag{1.5}$$

To check the result, substitute these resistances into Equations 1.1 and 1.2.

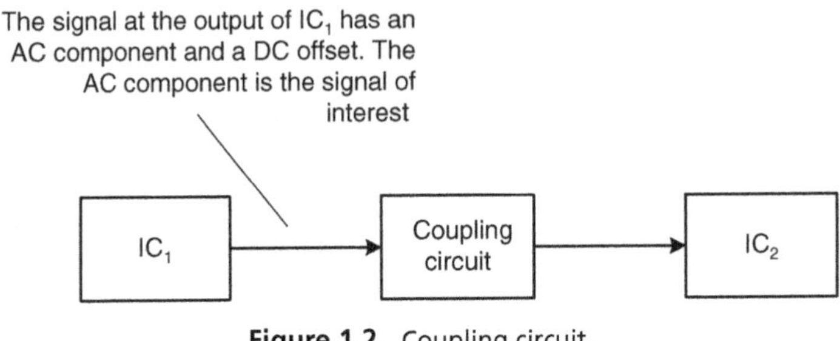

Figure 1.2. Coupling circuit.

1.2 DESIGN OF A COUPLING CIRCUIT

A common application for resistive circuits is the coupling circuit of Figure 1.2 that conditions a signal from integrated component IC_1 and feeds it to integrated component IC_2. The signal of interest, or the signal that carries useful information, is an alternating current (AC) signal with no DC component. However, DC offsets play an important role in this application because they keep the input and output signals within the working range of the amplifiers.

The required functionality of this circuit is to

1. Provide a specified load impedance for IC_1.
2. Attenuate the signal from IC_1 to IC_2 by a specified amount.
3. Provide a specified DC offset at the input of IC_2.

This functionality is provided by the circuit of Figure 1.3. Our design strategy is to first analyze the circuit and then use the analysis results to compute the component values.

The focus of this chapter is resistive circuits, so the capacitor in Figure 1.3 deserves explanation. It is called a "blocking capacitor" because its primary function is to prevent or block the DC voltage at the output of IC_1 so that it does not appear at the input of IC_2. The capacitor is selected so that it approximates a short circuit to the signal of interest. Chapter 3 shows how to select the correct value of this capacitor.

Prior to analyzing the circuit of Figure 1.3, we digress briefly to clarify the concept of "AC ground." This powerful tool is often a source of confusion for both students and experienced engineers. The confusion arises from not fully grasping the difference between *resistance* and *impedance*. *Resistance* is the ratio of the DC voltage across a device to the DC current passing through it. *Impedance* is the frequency-dependent, complex ratio of the AC voltage to

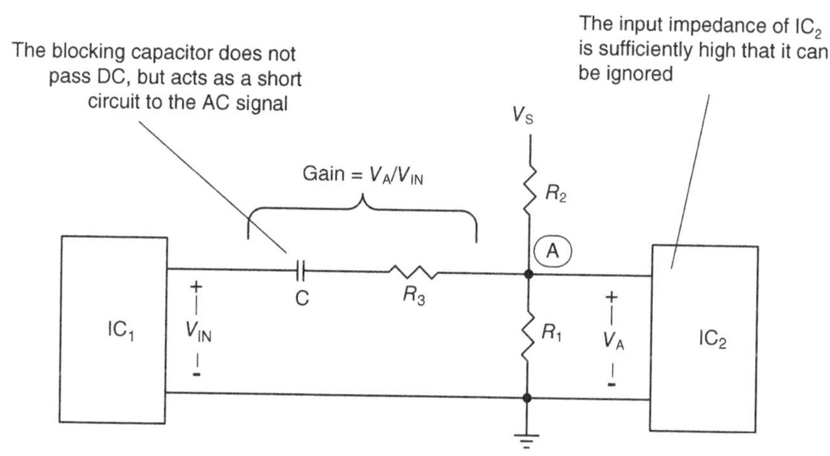

Figure 1.3. This coupling circuit provides a specified load impedance for IC_1, attenuates the signal from IC_1 to IC_2, and provides a specified DC bias for IC_2.

the AC current in a device or circuit. One way to remember this is to think of impedance as the derivative:

$$\text{Impedance} = Z = \frac{\partial V}{\partial I} \tag{1.6}$$

Note from Equation 1.6 that the impedance of a resistor is equal to its resistance.

Now consider an ideal DC power supply. Its function is to provide a fixed voltage regardless of the load current. Since its voltage does not change, Equation 1.6 shows that its impedance is zero.[2] Therefore, when computing the impedance of the circuit in Figure 1.3, we treat the power supply, V_S, as an AC ground.

Similarly the *attenuation* provided by the circuit of Figure 1.3 refers to the ratio of the AC voltage at the output of IC_1 to the AC voltage at A. When computing attenuation, the power supply, V_S, is also an AC ground.

The preceding discussion shows that for computing input impedance and gain,[3] the DC supply, V_S, is an AC ground which places R_1 in parallel with R_2 in Figure 1.3. We therefore simplify Figure 1.3 by representing R_1, R_2, and V_S as a DC Thevenin source as shown on the right side of Figure 1.4a. Since the capacitor appears as a short circuit to the desired signal and since the Thevenin voltage source is an AC ground, the circuit of Figure 1.4b can be used to compute the input impedance and gain.

[2] A large capacitor connected from a circuit node to ground results in an AC ground also.

[3] For the remainder of this discussion we use gain, V_A/V_{IN}, which is the inverse of attenuation.

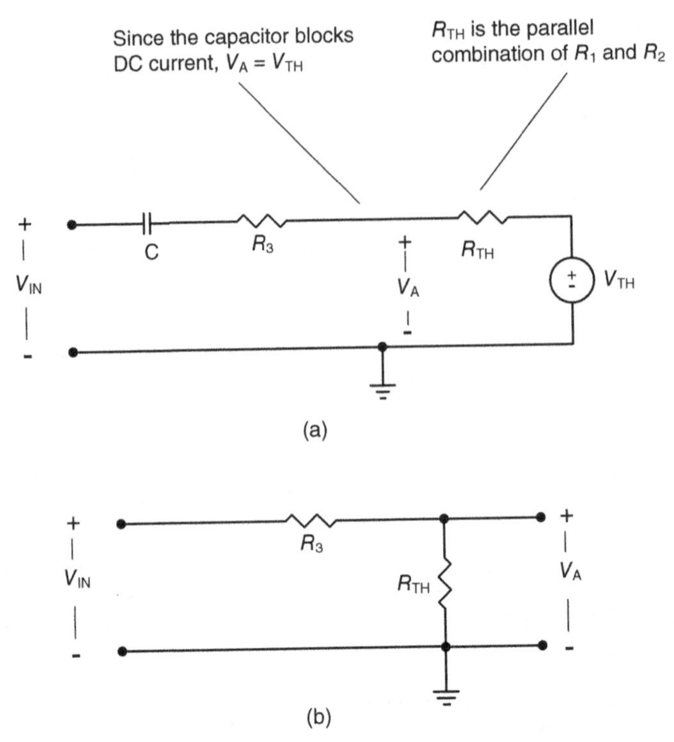

Figure 1.4. Simplifications of the circuit of Figure 1.3 are used to compute input impedance and circuit gain. A Thevenin equivalent allows R_1, R_2, and V_S to be shown as a Thevenin equivalent (a). Since the capacitor is an AC short and since voltage source V_{TH} is an AC ground, the circuit can be further simplified (b).

From Figure 1.4b the input impedance is

$$Z_{IN} = R_3 + R_{TH} \tag{1.7}$$

And the gain is computed using the voltage divider theorem

$$\text{Gain} = \frac{V_A}{V_{IN}} = \frac{R_{TH}}{R_3 + R_{TH}} \tag{1.8}$$

The analysis above can be used to design the coupler based on a design specification as shown in the following example.

Example 1.2. The circuit shown in Figure 1.5 is a coupling circuit for an AC signal. Assume that the amplifier input impedance is infinite, and the capacitor acts as a short circuit to the input signal. Pick resistors R_1, R_2, and R_3 so the DC bias at the amplifier input is 4 VDC, the signal from V_{IN} to the

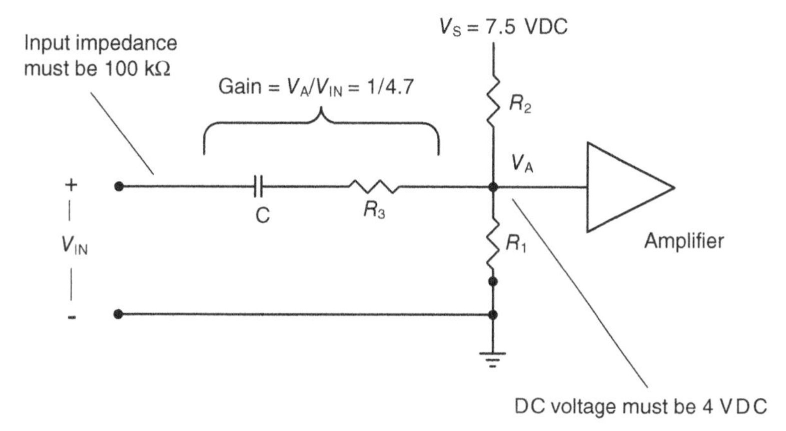

Figure 1.5. Circuit for Example 1.2.

amplifier input is attenuated by a factor of 4.7, and the impedance seen by the input signal at V_{IN} is 100 kΩ.

Solution. From the analysis above, we use the circuit of Figure 1.6 to compute input impedance and gain.

Substituting Z_{IN} from Equation 1.7 into Equation 1.8 and solving for R_{TH} gives

$$R_{TH} = \text{Gain} \times Z_{IN} = \frac{1}{4.7} 100\,\text{k}\Omega = 21.28\ \text{k}\Omega \tag{1.9}$$

This allows us to compute R_3 using Equation 1.7:

$$R_3 = Z_{IN} - R_{TH} = 100\,\text{k}\Omega - 21.28\,\text{k}\Omega = 78.72\ \text{k}\Omega \tag{1.10}$$

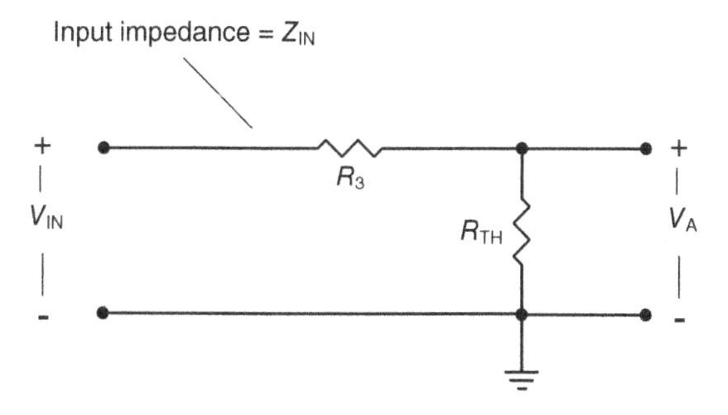

Figure 1.6. Thevenin equivalent for computing impedance and gain for Example 1.2.

Since the capacitor blocks DC current, the Thevenin voltage is the DC bias voltage of 4 VDC at the amplifier input. Knowing the Thevenin voltage, Thevenin resistance, and supply voltage, V_S, we compute R_1 and R_2 using Equations 1.5 and 1.4, respectively. From Equation 1.5:

$$R_1 = \frac{R_{TH}}{\left(1 - \frac{V_{TH}}{V_S}\right)} = \frac{21{,}280}{\left(1 - \frac{4}{7.5}\right)} = 45.60\ \text{k}\Omega \tag{1.11}$$

From Equation 1.4:

$$R_2 = \frac{R_{TH} V_S}{V_{TH}} = \frac{21{,}280 \times 7.5}{4} = 39.90\ \text{k}\Omega \tag{1.12}$$

1.3 DESIGN OF A PI ATTENUATOR

In this section we again begin with a design requirement, write descriptive equations for the proposed circuit, and then use the equations to compute the component values. We also show how mesh analysis and a matrix solver can be used to quickly and efficiently check our work by analyzing the resulting circuit.

The Pi network of Figure 1.7 is commonly found in radio frequency (RF) circuits and does the following:

1. Attenuates bidirectionally. The voltage gain, A_V, from left to right is identical to the voltage gain from right to left.
2. Provides a specified load resistance to the circuit that drives it.
3. Provides a specified source resistance to the circuit that it drives.

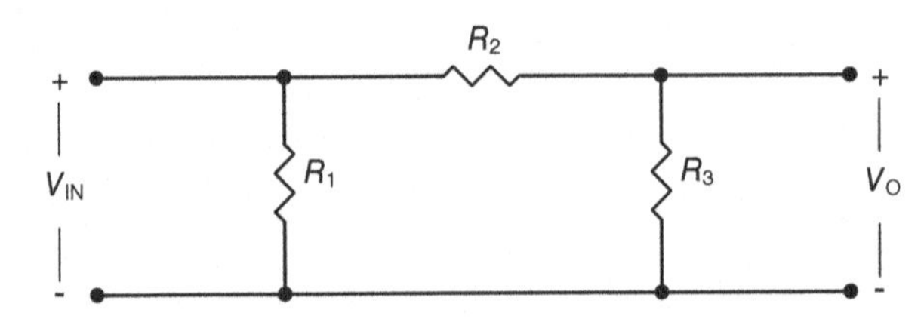

Figure 1.7. Pi network. This network is frequently used as a constant impedance RF attenuator.

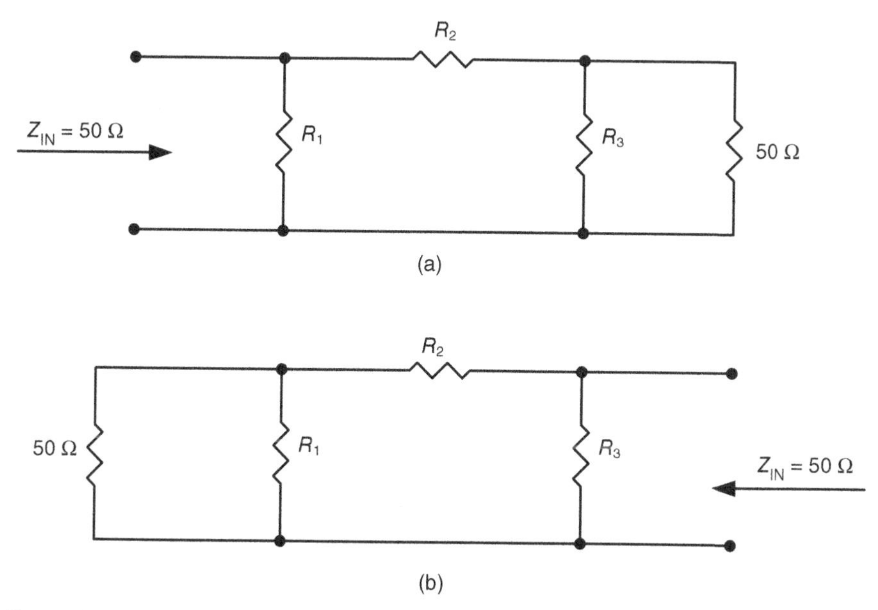

Figure 1.8. Pi network. With the output terminated in 50 Ω the input impedance is 50 Ω (a). With the input terminated in 50 Ω the output impedance is 50 Ω (b).

This network is important to RF designers because most RF components are designed to work with 50 Ω input and output impedances.[4] Of course, there are many online calculators such as [1] available for this purpose. But the skills used in this example will help you work with any network you'll encounter in an interview or on the job.

Specific design requirements for a 3 dB Pi attenuator are as follows:

1. With the output terminated in 50 Ω, the input impedance must be 50 Ω.
2. With the input driven by a 50 Ω source and the output terminated in 50 Ω, the ratio V_O/V_{IN} must be A_V.
3. With the input driven by a 50 Ω source, the output impedance must be 50 Ω.

The toughest part of this problem is to convert the design requirements into solvable equations. First, note that the problem can be simplified using symmetry as shown in Figure 1.8. When the *output* is terminated in 50 Ω the resulting circuit is simply the reflection of the circuit when the *input* is terminated in 50 Ω. Therefore, R_1 and R_3 are identical, and the number of

[4]See Chapter 9 for examples using attenuators.

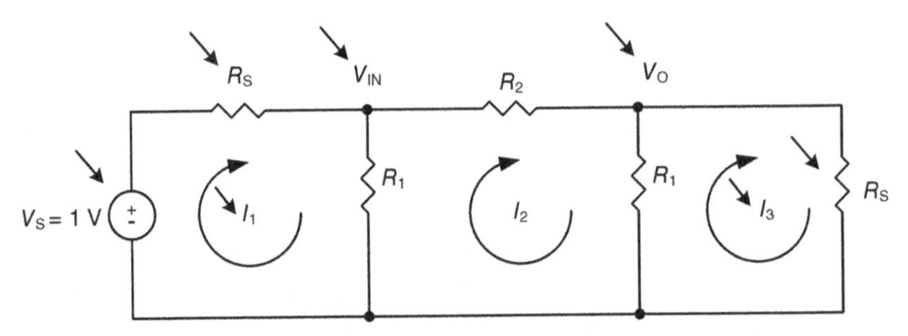

Figure 1.9. Pi network analysis model. The arrows point to quantities that are either given or can be readily computed by inspection. The unknowns in this problem are R_1, R_2, and I_2. $R_S = 50\ \Omega$.

components which must be determined is lowered from three to two. For our analysis R_3 is replaced by R_1, as shown in Figure 1.9.

The model of Figure 1.9 is used to describe the voltages and currents in the system. The calculations are simplified by setting $V_S = 1$ V. From requirement 1, the input resistance of the terminated network must be 50 Ω. Since R_S is 50 Ω voltage division gives

$$V_{IN} = \frac{V_S}{2} = \frac{1}{2} \tag{1.13}$$

Similarly, since the resistance seen by V_S is twice R_S or 100 Ω

$$I_1 = \frac{1}{2R_S} \tag{1.14}$$

Requirement 2 specifies that the ratio V_O/V_{IN} must be A_V. With voltage gain A_V known, V_O and I_3 can be computed:

$$V_O = V_{IN}A_V = \tfrac{1}{2}A_V \tag{1.15}$$

$$I_3 = \frac{V_O}{R_S} = \frac{A_V}{2R_S} \tag{1.16}$$

Returning to Figure 1.9, V_S, R_S, I_1, V_{IN}, I_3, and V_O are known and marked with arrows. R_1, R_2, and I_2 are unknown. Since there are three unknowns, three equations are needed. The first equation is written by observing that V_{IN} and I_1 are known and related to R_1 and I_2:

$$\frac{V_{IN}}{R_1} = I_1 - I_2 \tag{1.17}$$

The second equation is written by observing that V_{IN} and V_O are known and related to R_2 and I_2:

$$V_{IN} - I_2 R_2 = V_O \tag{1.18}$$

The third equation is written by observing that V_O and I_3 are known and related to R_1 and I_2:

$$\frac{V_O}{R_1} = I_2 - I_3 \tag{1.19}$$

Equations 1.17, 1.18, and 1.19 represent three nonlinear equations in three unknowns. These equations can be solved "by hand" as shown below or with a symbolic equation solver such as the one provided by Wolfram Mathematica [2].

We solve for the unknowns below using the following strategy:

1. Solve Equation 1.17 for R_1 as a function of I_2.
2. Substitute the expression for R_1 into Equation 1.19. Solve for I_2.
3. Substitute the value of I_2 into Equation 1.18. Compute R_2.
4. Substitute the value of I_2 into Equation 1.17. Compute R_1.

Equation 1.17 is solved for R_1:

$$R_1 = \frac{V_{IN}}{I_1 - I_2} \tag{1.20}$$

Equation 1.20 is substituted for R_1 in Equation 1.19 and then Equation 1.19 is solved for I_2:

$$I_2 = \frac{\frac{V_O}{V_{IN}} I_1 + I_3}{1 + \frac{V_O}{V_{IN}}} \tag{1.21}$$

Substitution of A_V for V_O/V_{IN}, Equation 1.14 for I_1, and Equation 1.16 for I_3 into Equation 1.21 gives

$$I_2 = \frac{A_V}{R_S\left(1 + A_V\right)} \tag{1.22}$$

Equation 1.22 is substituted into Equation 1.18 and Equation 1.18 is solved for R_2:

$$R_2 = \frac{\left(V_{\mathrm{IN}} - V_{\mathrm{O}}\right)\left(A_{\mathrm{V}} + 1\right) R_{\mathrm{S}}}{A_{\mathrm{V}}} \tag{1.23}$$

Substituting Equation 1.15 for V_{O} and Equation 1.13 for V_{IN} into Equation 1.23 gives

$$R_2 = \frac{\left(1 - A_{\mathrm{V}}^2\right)}{2A_{\mathrm{V}}} R_{\mathrm{S}} \tag{1.24}$$

Finally, substitution of Equation 1.13 for V_{IN}, Equation 1.14 for I_1, and Equation 1.22 for I_2 into Equation 1.17 gives

$$R_1 = \left[\frac{1 + A_{\mathrm{V}}}{1 - A_{\mathrm{V}}}\right] R_{\mathrm{S}} \tag{1.25}$$

Example 1.3. Design a 3 dB, 50 Ω, Pi attenuator. Verify the design gives the specified impedance and attenuation.

Solution. The attenuation is specified as 3 dB. This is converted to voltage gain A_{V} using the definition of the decibel[5]

$$\mathrm{dB} = 20 \log_{10}(A_{\mathrm{V}}) \tag{1.26}$$

Solving for voltage gain A_{V} gives[6]

$$A_{\mathrm{V}} = 10^{-\mathrm{dB}/20} = 10^{-3/20} = 0.708 \tag{1.27}$$

Knowing A_{V} and R_{S}, we use Equations 1.24 and 1.25 to compute the resistors as shown below:

$$R_2 = \frac{\left(1 - A_{\mathrm{V}}^2\right)}{2A_{\mathrm{V}}} R_{\mathrm{S}} = \frac{\left(1 - 0.708^2\right)}{2 \times 0.708} 50 = 17.6\ \Omega \tag{1.28}$$

[5]Note that the attenuation is specified as 3 dB, not 3 dB *voltage* or *power* gain. Students frequently get confused using dB on interviews. A given value of dB specifies *different* values for the voltage and power gains.

[6]The negative sign in front of dB in Equation 1.27 is used because attenuation has been specified as a positive quantity.

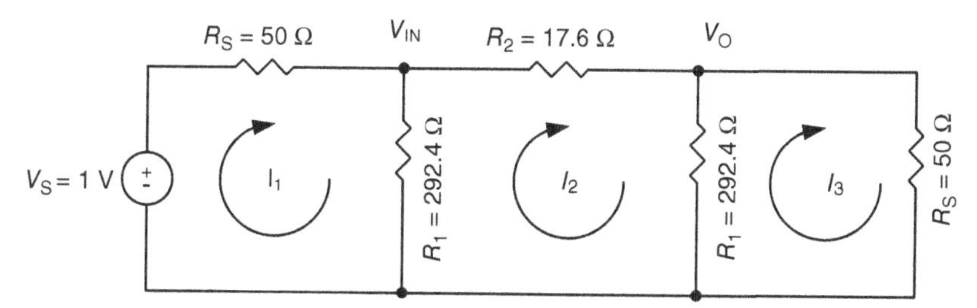

Figure 1.10. Resistor values for the Pi attenuator are checked by writing and solving mesh equations. Next, the input impedance and attenuation are compared to the specified values.

$$R_1 = \left[\frac{1 + A_V}{1 - A_V}\right] R_S = \left[\frac{1 + 0.708}{1 - 0.708}\right] 50 = 292.4\ \Omega \tag{1.29}$$

To check this result, the circuit of Figure 1.10 is analyzed by writing three equations in the three unknown mesh currents, and then solving them using $V_S = 1$ V.[7] We then verify that the input impedance is 50 Ω and the attenuation is 3 dB.

Equations for the three meshes are

$$\begin{aligned} -V_S + I_1 R_S + (I_1 - I_2) R_1 &= 0 \\ (R_S + R_1) I_1 - R_1 I_2 &= V_S \end{aligned} \tag{1.30}$$

$$\begin{aligned} (I_2 - I_1) R_1 + I_2 R_2 + (I_2 - I_3) R_1 &= 0 \\ -R_1 I_1 + (2R_1 + R_2) I_2 - R_1 I_3 &= 0 \end{aligned} \tag{1.31}$$

$$\begin{aligned} (I_3 - I_2) R_1 + I_3 R_S &= 0 \\ -R_1 I_2 + (R_1 + R_S) I_3 &= 0 \end{aligned} \tag{1.32}$$

In matrix notation this is

$$[\boldsymbol{A}][\boldsymbol{I}] = [\boldsymbol{V}] \tag{1.33}$$

[7]Chapter 3 shows that this technique is also used for AC circuits.

where

$$[A] = \begin{bmatrix} R_S + R_1 & -R_1 & 0 \\ -R_1 & 2R_1 + R_2 & -R_1 \\ 0 & -R_1 & R_1 + R_S \end{bmatrix} \tag{1.34}$$

$$[I] = \begin{bmatrix} I_1 \\ I_2 \\ I_3 \end{bmatrix} \tag{1.35}$$

$$[V] = \begin{bmatrix} V_S \\ 0 \\ 0 \end{bmatrix} = \begin{bmatrix} 1 \\ 0 \\ 0 \end{bmatrix} \tag{1.36}$$

To solve Equation 1.33 premultiply both sides by the inverse of $[A]$ or $[A]^{-1}$

$$[A]^{-1}\,[A]\,[I] = [A]^{-1}\,[V] \tag{1.37}$$

Since a matrix multiplied by its inverse gives the identity matrix, and multiplying a matrix by the identity matrix gives the original matrix,

$$[I] = [A]^{-1}\,[V] \tag{1.38}$$

Using an matrix tool such as [3] or Matlab gives $I_1 = 10$ mA, $I_2 = 8.29$ mA, and $I_3 = 7.08$ mA.

We check for 50 Ω input impedance by verifying that voltage V_{IN} is half the input voltage, V_S, or 0.5 V:

$$V_{IN} = 1 - I_1 \times 50 = 0.5 \text{ V} \tag{1.39}$$

We check for 3 dB attenuation by verifying that the output voltage, V_{OUT} is 0.7071 times the input voltage, V_{IN}.

$$V_O = I_3 R_S = 7.08 \times 10^{-3} \times 50 = 0.7071 \times 0.5 \tag{1.40}$$

PROBLEMS

1.1 Given a 5 V supply, design a circuit with $V_{TH} = 3.3$ V and $R_{TH} = 1000$ Ω.

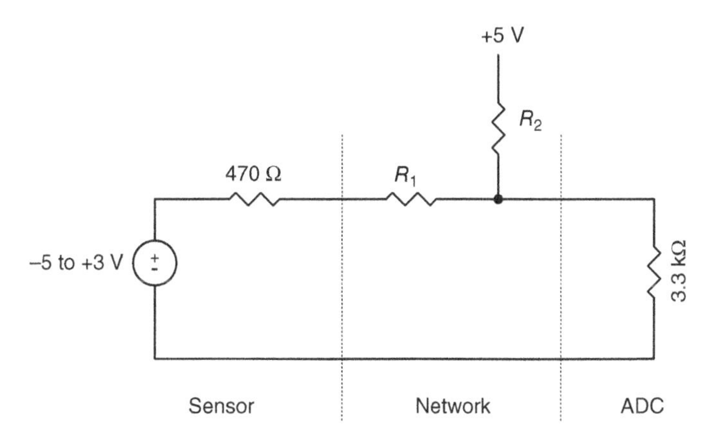

Figure 1.11. Circuit for Problem 1.2.

1.2 A sensor produces a voltage between −5 and +3 V. The output resistance of the sensor is 470 Ω. The sensor must be interfaced to an ADC with an input range of 0 to +3.3 V and an input resistance of 3300 Ω. Determine the values of R_1 and R_2 for the interface network shown in Figure 1.11. Check your work by analyzing the resulting network.

1.3 Components IC_1 and IC_2 are connected using the coupling network shown in Figure 1.12. IC_1 has a DC offset of 15 VDC and must be loaded with 41 kΩ. The AC signal from IC_1 is 7 Vpp. IC_2 requires a DC bias of 4.2 VDC and a input signal level of 3 Vpp. The input of IC_2 has a 47 kΩ resistor connected directly to ground. Compute R_1, R_2, and R_3 for this network.

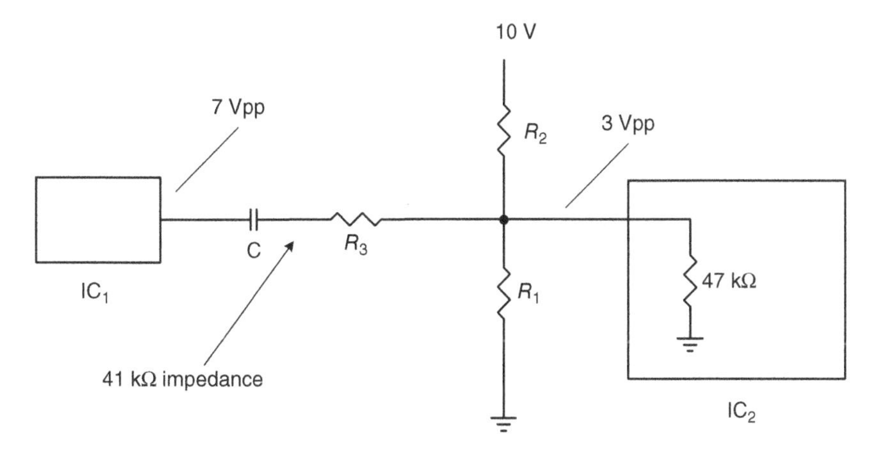

Figure 1.12. Circuit for Problem 1.3.

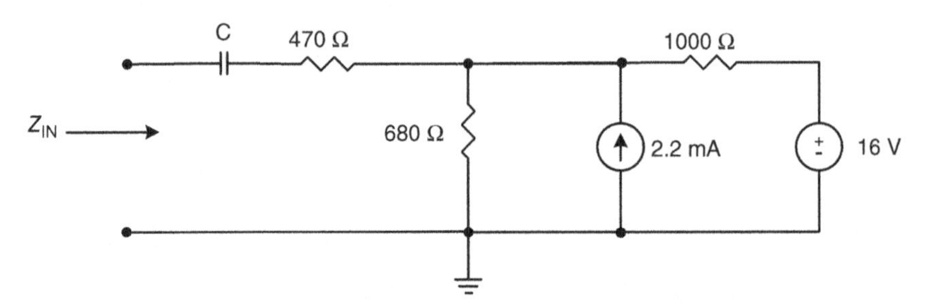

Figure 1.13. Circuit for Problem 1.5.

1.4 Using an argument similar to the one used for a DC voltage source in Section 1.2, determine the impedance of a DC current source. How should a DC current source be represented when analyzing an AC circuit? *Hint:* Use Equation 1.6.

1.5 Compute the input impedance of the circuit of Figure 1.13 for AC signals. Assume the frequency is sufficiently high so the capacitor can be considered a short circuit.

1.6 A coupling network connects components IC_1 and IC_2 as shown in Figure 1.14. Assume the output impedance of IC_1 is zero, the input impedance of IC_2 is infinite, and the capacitor appears as a short circuit to the signal of interest.

(a) Determine the attenuation provided by the network.

(b) Determine the DC bias at the input of IC_2.

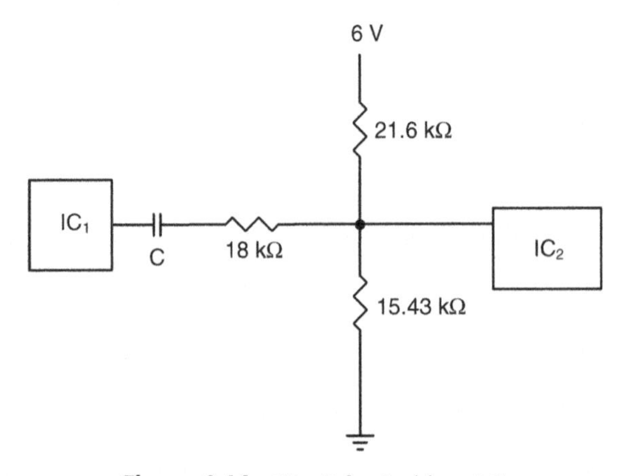

Figure 1.14. Circuit for Problem 1.6.

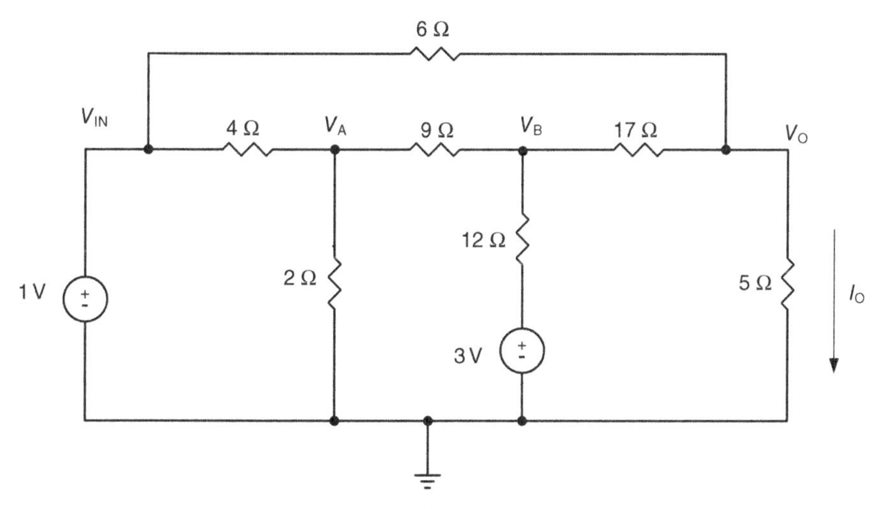

Figure 1.15. Circuit for Problems 1.7 and 1.8.

(c) Determine the load impedance seen by IC_1.
(d) Determine the source impedance seen by IC_2. *Hint:* It is the impedance looking from IC_2 back to the network.

1.7 Determine current I_O in the circuit of Figure 1.15 using mesh analysis.

1.8 Determine total power dissipation in the circuit of Figure 1.15.

REFERENCES

1. Microwaves101. *Attenuator Calculator*. Available: http://www.microwaves101.com/encyclopedia/calcattenuator.cfm (Accessed 11-29-2013.)
2. W. Research. *Wolfram Mathematica*. Available: http://www.wolfram.com/mathematica/ (Accessed 11-29-2013.)
3. Bluebit. *Linear Equation Solver*. Available: http://www.bluebit.gr/matrix-calculator/linear_equations.aspx (Accessed 11-29-2013.).

2

HOW TO PREVENT A POWER TRANSISTOR FROM OVERHEATING

In addition to meeting electrical requirements, a successful circuit design must meet thermal requirements. This insures that the circuit will not fail prematurely due to excessive temperature and that it will not excessively raise the temperature of neighboring circuits and cause them to fail. If thermal analysis is not included in the overall system design, the individual circuits may work properly on the bench but fail when they are placed together in an enclosure. This chapter shows a fast and simple way to estimate component temperatures. It also shows how to specify a heat sink that will keep the temperature of a device at a safe level. This technique will not give the precise results obtainable using sophisticated thermal modeling tools, but it will allow you to quickly make rough thermal estimates as you consider different approaches during the early phase of a design.

Most experienced engineers have seen good electrical designs get scrapped because they were thermally impractical, so it is common for them to ask candidates to solve simple heat transfer problems during interviews. This chapter will reinforce your intuitive understanding of heat transfer, show you

Ten Essential Skills for Electrical Engineers, First Edition. Barry L. Dorr.

how to apply it in the workplace, and show examples of problems typically asked during interviews.

We begin by reviewing steady-state heat transfer problems. Next, we show how to use data provided by manufacturers of electronic components and heat sinks to keep electronic components at safe temperatures. The chapter concludes by showing how to compute device temperature as a function of time using the notion of thermal capacitance.

2.1 ELECTRICAL MODEL FOR HEAT TRANSFER

No one likes to burn their fingers, so most of us have developed good intuition about heat transfer. Consider what happens when the end of a metal rod is placed in a vat of hot liquid as shown in Figure 2.1. We intuitively expect that the part of the rod nearest to the liquid will be nearly as hot as the liquid

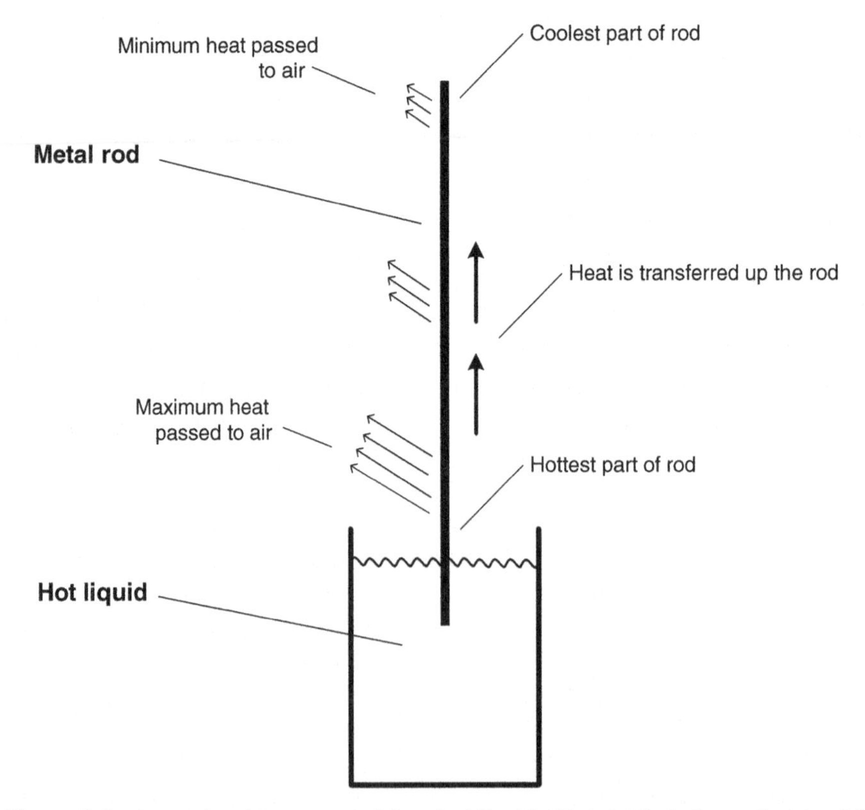

Figure 2.1. A metal rod is suspended in a hot liquid. We intuitively know that heat is transferred up the rod and from the rod to the surrounding air.

	Basic stuff	Flow of stuff	What pushes the stuff	What resists the flow of stuff	Storage of stuff
Electrical	Charge (Coulombs)	Current (Coulombs/second)	Voltage Difference (Joules/Coulomb)	Resistance (Ω)	Capacitance (Farads or Coulombs/V)
Thermal	Thermal Energy or Heat (Joules)	Power (Joules/second or Watts)	Temperature Difference (°C)	Thermal Resistance (°C/W)	Heat Capacity (Joules/°C)

Figure 2.2. The "stuff" flowing in a thermal system is heat—the "stuff" in an electrical system is charge.

and the temperature of the rod will decrease as the distance from the liquid increases. We also expect that the rod will pass more heat to the air at its hottest point and less at the end where it is coolest. Finally, we know that thermal energy from the hot liquid is being transferred from the liquid, and this will tend to cool the liquid.

As electrical engineers we observe that heat is flowing, and greater temperature differences are associated with more heat flow. We see that *temperature* pushes *heat* similar to the way *voltage* pushes *current*. We also observe that as heat flows from the hot liquid the temperature of the liquid drops which reminds us of the way the voltage on a capacitor drops as its charge is depleted. If the system of Figure 2.1 could be modeled with electrical components then circuit analysis tools could be used to analyze the system. Fortunately, this is indeed possible by relating thermal *heat* to electrical *charge*. The basic relationships are shown in Figure 2.2.

For electrical circuits Ohm's law states that the voltage rise across a resistive device is equal to the current passing through it multiplied by the resistance. Using the thermal analogies in Figure 2.2 we can write the thermal equivalent of Ohm's law:

$$\Delta T = P(\text{watts})\,\theta\left(\frac{°\text{C}}{\text{watt}}\right) \tag{2.1}$$

where ΔT is the temperature rise in degrees centigrade, P is the power in watts, and θ is the thermal resistance in °C/W.

Our modeling strategy is therefore to

1. Model a source of heat with an electrical current source. The value of current represents the power of the source in watts.
2. Model anything that resists the flow of heat with an electrical resistor. The value of resistance is equal to the thermal resistance.

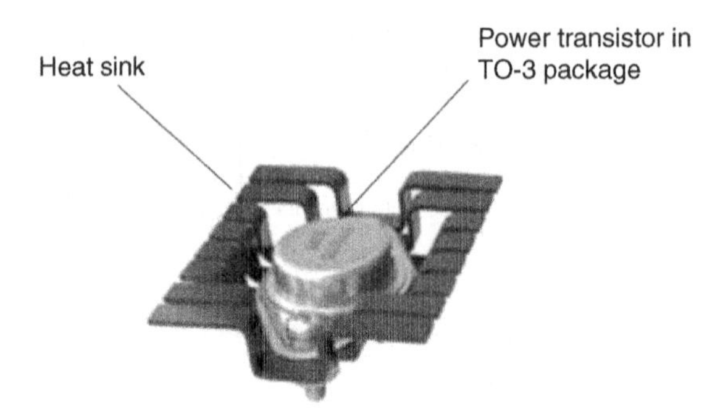

Figure 2.3. Transistor and heat sink for Example 2.1. (Courtesy of Aavid Corporation)

3. Model ambient thermal nodes, that is, nodes where the temperature remains the same when heat flows into or out of them, with an electrical voltage source. The voltage is equal to the temperature in degrees.[1]
4. Model a hot object with an electrical capacitor. The value of capacitance is equal to the heat capacity of the object.

This strategy is remarkably easy to use, as shown in the example below.

Example 2.1. A 2N3055 transistor in an industry standard TO-3 package is mounted to a heat sink as shown in Figure 2.3. The transistor dissipates 10 W and the ambient temperature is 20°C. Determine the junction temperature of the transistor and the temperature of the transistor case.

Solution. The model is shown in Figure 2.4. The transistor generates 10 W at its junction and is modeled with a 10 A current source. The heat is then transferred to the transistor's case through thermal resistance R_{JC} which accounts for the fact that the semiconductor junction is hotter than the case. The datasheet for the transistor gives R_{JC} as 1.5°C/W. The transistor is securely mounted to the heat sink[2] insuring the temperature at the case matches the temperature at the mating surface of the heat sink. The datasheet for the heat sink gives the thermal resistance from the mating surface of the heat sink to ambient, R_{SA}, as 7°C/W. The temperature of the air does not increase as it absorbs heat,[3] which allows us to model ambient air as a voltage source.

[1]We use the surrounding air as an ambient node in our discussion.

[2]In practice an electrically insulating but thermally conductive pad is often placed between the transistor and the heat sink. The thermal resistance of a pad used for a TO-3 case is about 0.5 °C/W.

[3]Beware; this assumption is often invalid if the heat sink is in a closed box!

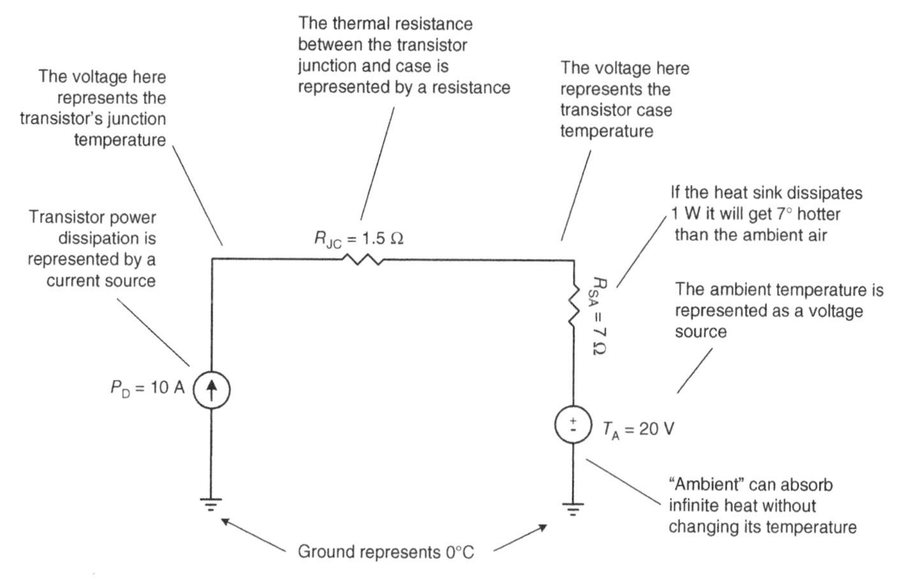

Figure 2.4. Model of transistor and heat sink for Example 2.1.

The temperature at any point in the system is the corresponding voltage in the circuit model. The transistor junction temperature is computed by starting at the voltage source and using Ohm's law:

$$T_J = T_A + P_D\left(R_{SA} + R_{JC}\right) = 20 + 10(7 + 1.5) = 105°\text{C} \tag{2.2}$$

We then compute the transistor case temperature as[4]

$$T_{CASE} = T_A + P_D R_{SA} = 20 + 10 \times 7 = 90°\text{C} \tag{2.3}$$

2.2 USING MANUFACTURER'S DATA FOR THERMAL ANALYSIS

Device manufacturers support the concept of electrical analogies for thermal analysis by providing thermal information in device datasheets as shown in Figure 2.5.

[4]Equations 2.2 and 2.3 point out that for constant power dissipation, the temperature of the system components track the ambient temperature.

Table 1. Maximum Ratings

Rating		Symbol	Value	Unit
Drain-Source Voltage		V_{DSS}	–0.5, +70	Vdc
Gate - Source Voltage		V_{GS}	–0.5, +15	Vdc
Total Device Dissipation @ T_C = 25°C Derate above 25°C	MRF373ALR1 MRF373ALSR1	P_D	197 1.12 278 1.59	W W/°C W W/°C
Storage Temperature Range		T_{stg}	–65 to +150	°C
Case Operating Temperature		T_C	150	°C
Operating Junction Temperature		T_J	200	°C

Table 2. Thermal Characteriatics

Characteristic		Symbol	Value	Unit
Thermal Resistance, Junction to Case	MRF373ALR1 MRF373ALSR1	$R_{\theta JC}$	0.89 0.63	°C/W

Figure 2.5. The datasheet for the Freescale MRF373A transistor provides data for thermal analysis. (Copyright 2013 Freescale Inc., used with permission)

Transistors sustain thermal damage when the junction temperature rises above the maximum value shown in the datasheet.[5] The junction temperature can be determined using Equation 2.1 and the model of Figure 2.4 but it is common for manufacturers to provide derating information so designers can easily determine the maximum power dissipation given the case temperature of the transistor. Figure 2.5 shows that the maximum power dissipation for the MRF373ALR1 transistor is 197 W and the derating factor is 1.12 W/°C. Limiting the power dissipation to the derated value will prevent the junction from exceeding the maximum value (see Problem 2.3). The transistor is derated using the equation

$$P_D = 197 - 1.12\left(T_C - 25\right) \tag{2.4}$$

where P_D is the maximum power dissipation at case temperature T_C degrees Celsius.

Instead of providing a derating factor, manufacturers sometimes provide a derating curve. The curve for the MRF373ALR1 transistor is drawn using Equation 2.4 and is shown in Figure 2.6.

Example 2.2. A MRF373ALR1 transistor must be mounted on a heat sink so that it can safely dissipate 20 W. The ambient air temperature is 30°C.

[5]The mean time before failure (MTBF) of a transistor decreases when it is operated near its maximum junction temperature.

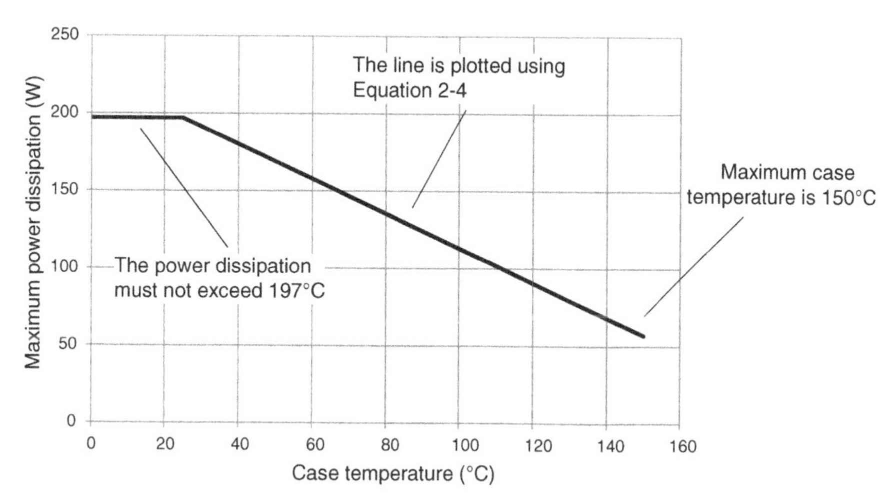

Figure 2.6. Derating curve for the MRF373ALR1 transistor. The safe area of operation is below the curve.

Choose a heat sink that will limit the junction temperature to the conservative value of 125°C. Show the operating point on the transistor's derating curve.

Solution. We use the electrical model shown in Figure 2.7 to compute the required thermal resistance for the heat sink. Equation 2.2 gives

$$T_J = 125 = T_A + P_D\left(R_{SA} + R_{JC}\right) = 30 + 20\left(R_{SA} + 0.89\right) \qquad (2.5)$$

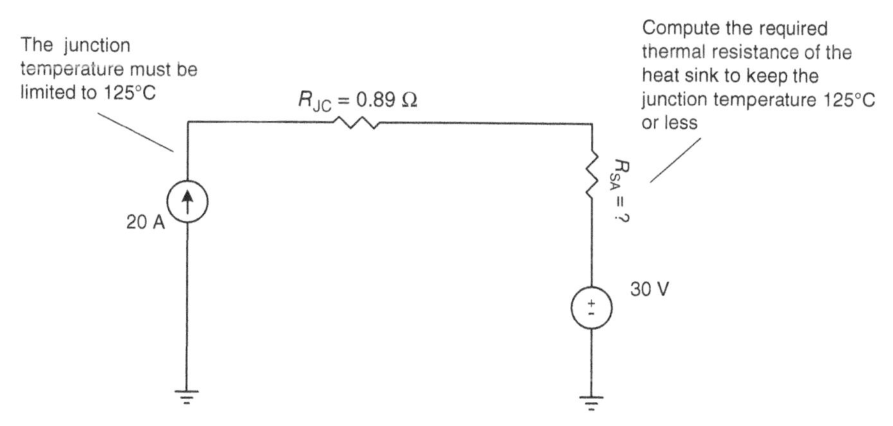

Figure 2.7. Electrical model for Example 2.2 used for computing the required thermal resistance of the heat sink.

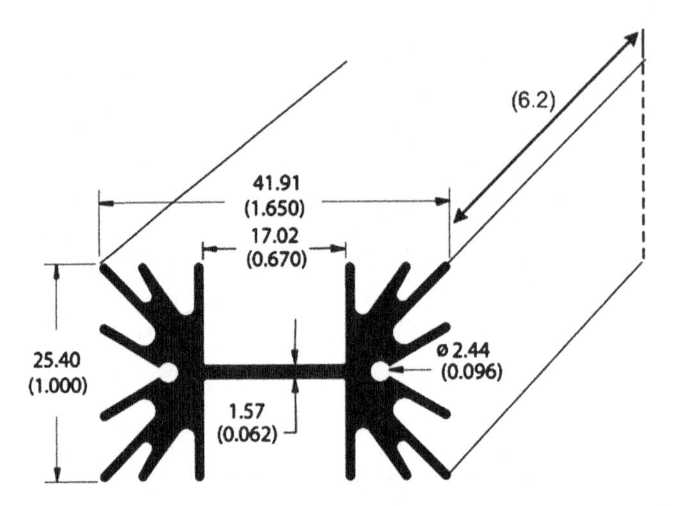

Figure 2.8. A 6.2-in. length of Aavid's 63130 extrusion provides the required thermal resistance for Example 2.2. The dimensions in parenthesis are inches. (Courtesy of Aavid Corporation)

Solving this equation gives the required thermal resistance of the heat sink. $R_{SA} = 3.86°C/W$.

Off-the-shelf heat sinks such as the one shown in Figure 2.3 are inexpensive and suitable for many applications. However, heat sinks can also be procured by specifying a length of extruded material. Manufacturers typically provide online calculators or equations for determining the length and Aavid's calculator shows that 6.2 in. of Aavid's 63130 extrusion [1] as shown in Figure 2.8 will provide the required thermal resistance.

Finally, we need to show the operating point of our system on the derating curve. The model of Figure 2.7 is used to solve for the case temperature:

$$T_C = 30 + 3.86 \times 20 = 107.2°C \tag{2.6}$$

The power dissipation and case temperature are shown on the derating curve of Figure 2.9.

2.3 FORCED-AIR COOLING

The system of Example 2.2 dissipated 20 W and required a heat sink about the size of a desk stapler. As engineers we immediately wonder if a smaller heat sink could provide the same cooling. Intuition correctly tells us that blowing cool air over a hot object will tend to cool it. We can use this fact to reduce

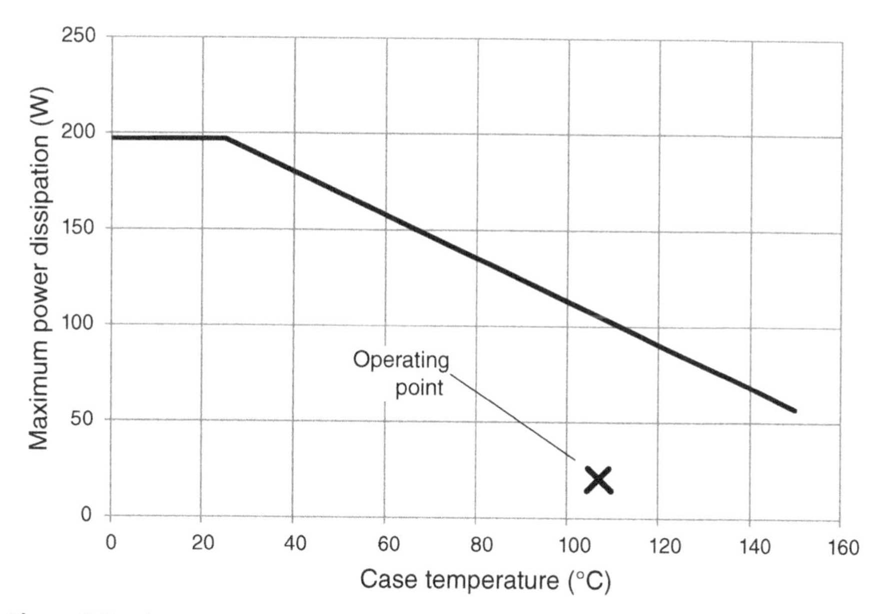

Figure 2.9. The operating point for the system in Example 2.2 is safely below the derating curve for the MRF373ALR1 transistor.

the size of the heat sink in Example 2.2. When air is blown over a heat sink its thermal resistance decreases. Therefore if forced air is available a smaller heat sink should suffice.[6]

Manufacturers of heat sinks provide data showing the effect of forced-air cooling. Figure 2.10 shows that if air at 200 ft/min is blown over the Aavid 63130 extrusion, the 3.86°C/W thermal resistance required for Example 2.2 can be obtained with a 1.75-in. section instead of a 6.2-in. section [1].

2.4 DYNAMIC RESPONSE OF A THERMAL SYSTEM

Until now we have discussed thermal systems with constant heat sources and we have computed the resulting steady-state temperatures. Real systems, however, are much more interesting—sources of heat are switched on and off, and power dissipation often varies with time. Now we extend our analysis to model dynamic conditions.

We know that placing an ice cube in a cup of hot coffee will lower the temperature of the coffee measurably. But if we place the ice cube in a swimming pool the change in the pool's temperature will not be noticeable.

[6] . . . but the fan better not fail. See Problem 2.4.

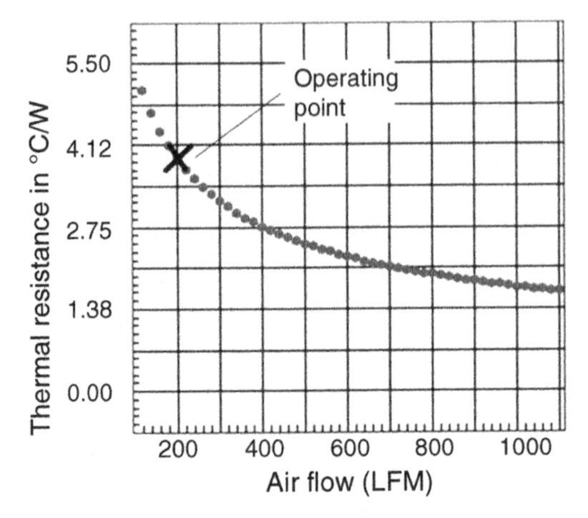

Figure 2.10. When air is blown over a 1.75-in. piece of Aavid 63130 extrusion at 200 ft/min the thermal resistance is the same as a 6.2-in. piece in still air. (Courtesy of Aavid Corporation)

In both cases the ice cube melts, meaning a fixed amount of thermal energy is transferred from the liquid to the ice cube. The swimming pool absorbs the energy with only a small decrease in temperature, but transferring the same amount of energy lowers the temperature of the coffee significantly. This is because the swimming pool has a higher *heat capacity* than the cup of coffee. Heat capacity is the amount of heat in joules required to change the temperature of an object by 1°C. Returning to Figure 2.2 we see that heat capacity is represented in our electrical model as capacitance.[7]

Including thermal capacitance in our models allows us to include the effect of time in our analysis. For example, we can gain insight into our design by computing the step response of the system when power is applied as shown in the following example.[8]

Example 2.3. Figure 2.11 shows a model of a transistor connected to a heat sink. The thermal resistance of the heat sink is 7 Ω and its heat capacity is 2 J/°C.[9] Initially all temperatures are 20°C. At time = 0^+ the transistor begins dissipating 10 W. Plot the transistor junction temperature as a function of time.

[7] Ambient has an infinite heat capacity.

[8] With a circuit model, a circuit simulator can be used to determine the thermal behavior for arbitrary inputs such as pulsed power sources. A circuit simulator also allows you to include your thermal system as part of a feedback loop.

[9] You can estimate the thermal capacitance experimentally. See Problem 2.6.

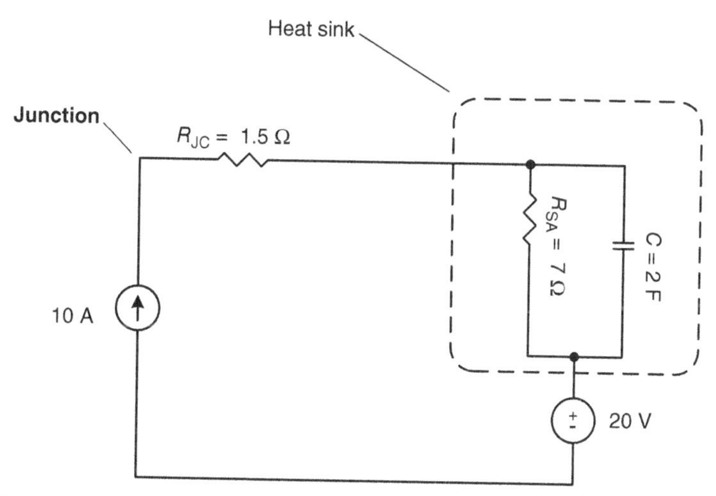

Figure 2.11. Model of a transistor and heat sink incorporating the thermal capacitance of the heat sink. The junction temperature rises exponentially when heat is applied.

Solution. For any RC circuit the response to a step input is

$$V(t) = V_F - \left(V_F - V_I\right) e^{-t/RC} \tag{2.7}$$

where V_F is the final voltage, V_I is the starting voltage, and R is the equivalent resistance seen by the capacitor.

Since the current source has infinite impedance, the only resistance seen by the capacitor is R_{SA} or 7 Ω. We compute V_I by noting that at time = 0^+ the capacitor appears as a short circuit so the initial voltage at the junction is computed using Ohm's law:

$$V_I = 20 + 10 \times 1.5 = 35 \text{ V} \tag{2.8}$$

At steady state, the capacitor appears as an open circuit and the final voltage at the junction is computed using Ohm's law:

$$V_F = 20 + 7 \times 10 + 1.5 \times 10 = 105 \text{ V} \tag{2.9}$$

The equation for the junction temperature versus time is written by substituting V_F, V_I, and R into Equation 2.7:

$$V(t) = 105 - (105 - 35)\, e^{-t/7\times 2} = 105 - 70e^{-t/14} \tag{2.10}$$

The junction temperature versus time is plotted in Figure 2.12.

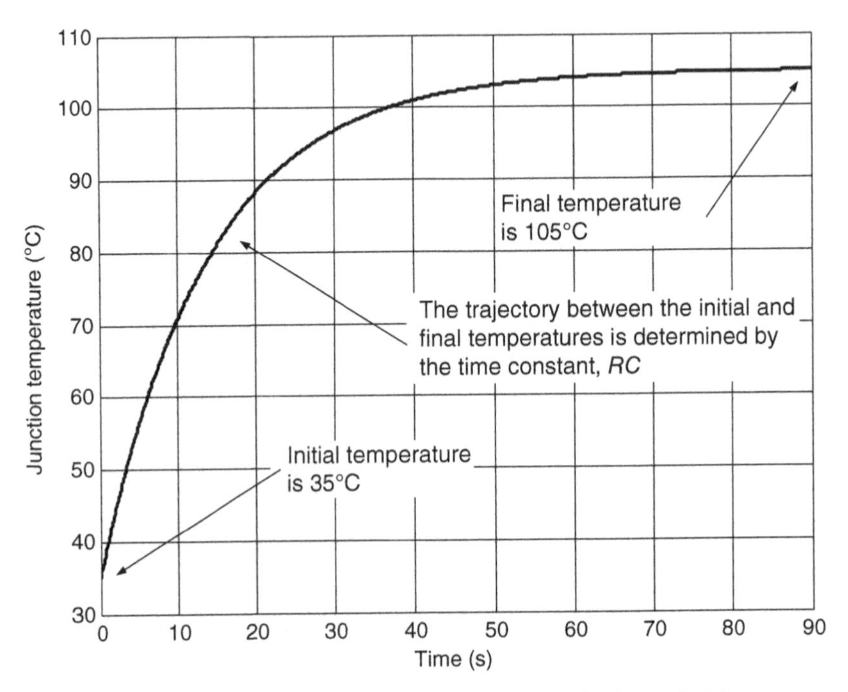

Figure 2.12. Junction temperature versus time for Example 2.3.

PROBLEMS

2.1 The maximum junction temperature for a transistor is 180°C. If the thermal resistance from junction to case is 0.88°C/W and the thermal resistance of the heat sink is 8°C/W, how much power can the transistor dissipate if the ambient temperature is 38°C?

2.2 A field effect transistor (FET) generates 22 W. $R_{JC} = 1$°C/W. The FET is attached to a heat sink with $R_{SA} = 10$°C/W. The FET also dissipates power from its case to the air. The thermal resistance from the case to ambient is $R_{CA} = 20$°C/W. The ambient air temperature is 27°C. Draw an electrical model for this situation and compute the transistor junction temperature.

2.3 Show that limiting the power dissipation of the MRF373ALR1 transistor to P_D from Equation 2.4 will cause the junction temperature to remain at 200.9°C regardless of the transistor case temperature. *Hint:* Use the equation $T_J = T_C + R_{JC}P_D$ and Equation 2.4.

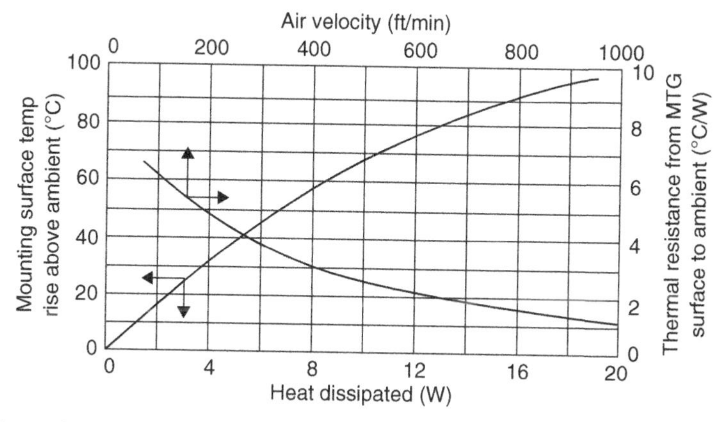

Figure 2.13. Heat sink data for Problem 2.5. (Courtesy of Aavid Corporation)

2.4 Using forced-air cooling allowed us to reduce the length of the heat sink in Example 2.2 from 6.2 to 1.75 in. If air is not blown over the heat sink the thermal resistance is 7.26°C/W. Compute the junction temperature if the cooling fan fails.

2.5 A transistor dissipating 9 W with $R_{JC} = 1.9$°C/W is mounted on the heat sink described by Figure 2.13. The ambient temperature is 40°C. Determine the minimum forced-air velocity required to keep the transistor junction temperature below 110°C.

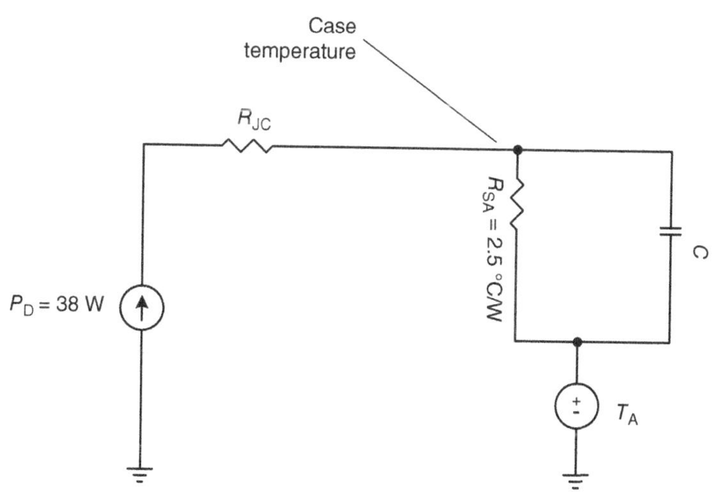

Figure 2.14. Circuit for Problem 2.6.

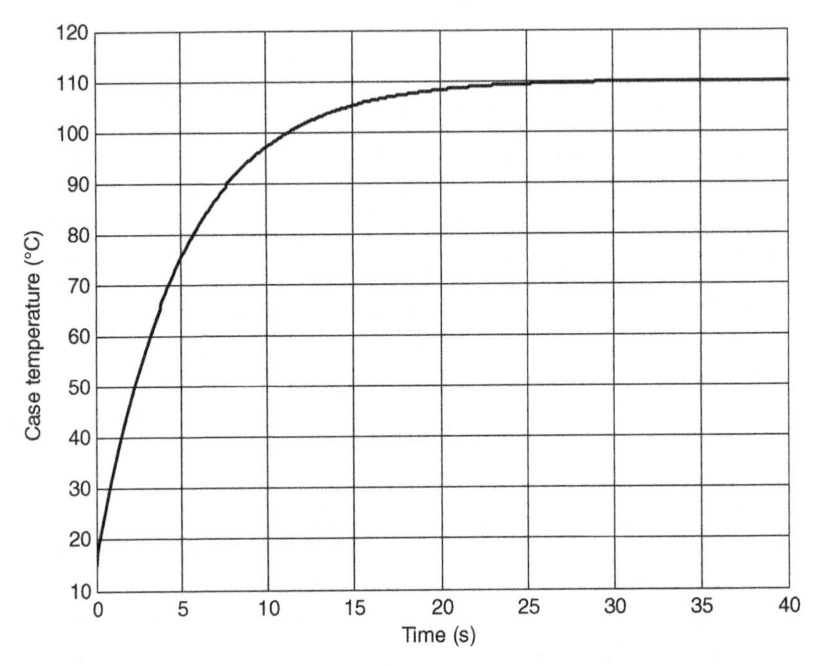

Figure 2.15. Temperature measured at transistor case for Problem 2.6.

2.6 A transistor is connected to a heat sink as shown in Figure 2.14. At time = 0^- all the system components are at the ambient temperature of 15°C. At time = 0^+ the transistor begins dissipating 38 W. The case temperature is measured and plotted versus time in Figure 2.15. Estimate the thermal capacity for the heat sink.

REFERENCE

1. Aavid Thermalloy. *Datasheet for Extrusion 63130*. Available: http://www.aavid.com/products/extrusion-heatsinks/63130 (Accessed 11-29-2013.)

3

HOW TO ANALYZE A CIRCUIT

Most of the analog circuits that you will encounter in the workplace and in job interviews will be fairly simple. The circuitry on most circuit boards consists of integrated components connected together by small "clumps" of interface circuitry that provide functionality such as level shifting, filtering, or signal conditioning. Frequently this circuitry is not documented, and, if you inherit the design from another engineer, you will need to figure out what it does and whether it does it correctly. This chapter shows techniques for quickly and accurately analyzing these kinds of circuits.

Analyzing simple circuits is a critical skill for job interviews, but you may question the value of this chapter when an inexpensive circuit simulator could be used to plot the frequency response for most networks. In addition to showing you how to quickly determine what a circuit does, this chapter will provide you with intuition and insight required for designing analog networks. By using the techniques reviewed here, you will not only get the analysis result, but add the circuit to your experience base and be able to

Ten Essential Skills for Electrical Engineers, First Edition. Barry L. Dorr.

modify and use it in the future. Good circuit designers always keep an arsenal of clever and useful circuits at their disposal.

Our approach to circuit analysis is as follows:

1. Simplify the circuit using techniques such as superposition and Thevenin's theorem.
2. Write equations that describe the system.
3. Check the equations at DC and infinite frequency using *asymptotic analysis.*
4. Use a tool such as Matlab to do the computations for the frequency response or impedance.

This chapter does not review methods for drawing Bode plots [1] because we can use the computer for plotting frequency response. However, note that most of the examples are reduced to the form used for drawing a Bode plot. If you are familiar with Bode analysis, you are encouraged to supplement the examples and problems in this chapter by drawing Bode plots as they provide exceptional insight into analog networks along with being extremely useful when designing circuits.

The chapter begins by using Matlab to plot the frequency response of an s-domain transfer function. This is followed by several examples of analyzing practical coupling circuits. We then show a powerful technique for analyzing *ladder networks*, which are commonly used for analog filters. Finally, we show how to analyze an arbitrary passive network using matrix techniques similar to those used in Chapter 1.

3.1 FREQUENCY RESPONSE OF A TRANSFER FUNCTION

Analog networks can be described by frequency-dependent transfer functions. The transfer function is the complex ratio of the output signal to the input signal. For example, the transfer function of a lowpass RC filter relates the output voltage to the input voltage:

$$\frac{V_O(s)}{V_{IN}(s)} = \frac{1}{RC}\frac{1}{s + \frac{1}{RC}} \tag{3.1}$$

To compute the frequency response substitute $s = jw = 2\pi f$ into Equation 3.1, where f is the frequency in hertz. This is done at a set of frequencies and the resulting set of complex values is the frequency response. Frequency

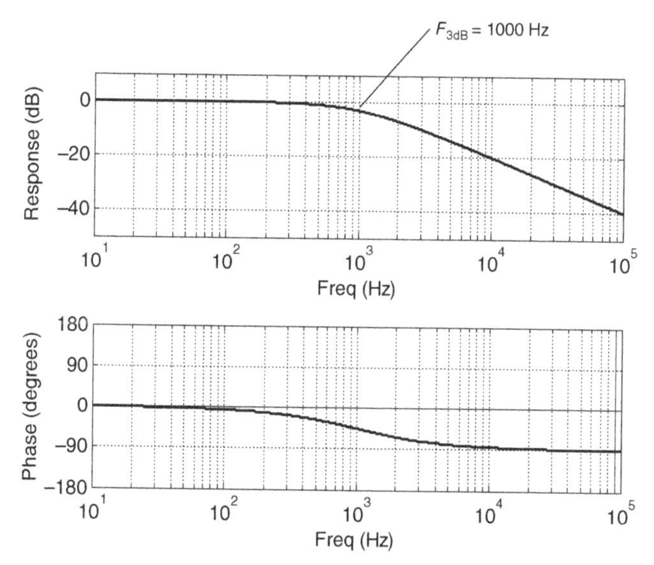

Figure 3.1. Frequency response plot for a lowpass RC filter with $F_{3dB} = 1000$ Hz.

response is typically displayed by showing the magnitude in decibels and the phase in radians or degrees as shown in Figure 3.1.

The process of determining the transfer function for a circuit often requires a significant amount of algebra and is consequently prone to errors. Fortunately, it is possible to partially check the transfer function by comparing it to the circuit and using the technique of *asymptotic analysis*. This does not guarantee that the transfer function is correct at all frequencies, but does verify that the response is correct at DC and at infinite frequency. Asymptotic analysis is a powerful tool that will almost always reveal errors in transfer functions.[1]

The impedances of circuit elements based on frequency are shown in Table 3.1.

Figure 3.2a shows a lowpass RC circuit and its transfer function. Figure 3.2b shows the equivalent circuit at DC where it is seen that the gain is unity. The equation below Figure 3.2b shows the substitution of $s = j0$ in the transfer function, which also gives unity. Figure 3.2c shows the equivalent circuit at very high frequency where it is seen that the gain is zero. The equation below Figure 3.2c shows that the substitution of $s = j\infty$ in the transfer function also gives zero.

[1]If you are asked to determine a transfer function on a job interview, always check your result using asymptotic analysis. If your transfer function is incorrect, but you can show it's incorrect with this technique, it demonstrates you have good understanding of circuit analysis.

TABLE 3.1. Impedances of circuit elements at different frequencies

	Resistor	Capacitor	Inductor
Arbitrary frequency	R	$\frac{1}{j2\pi fC}$	$j2\pi fL$
$f = 0$	R	∞	0
$f = \infty$	R	0	∞

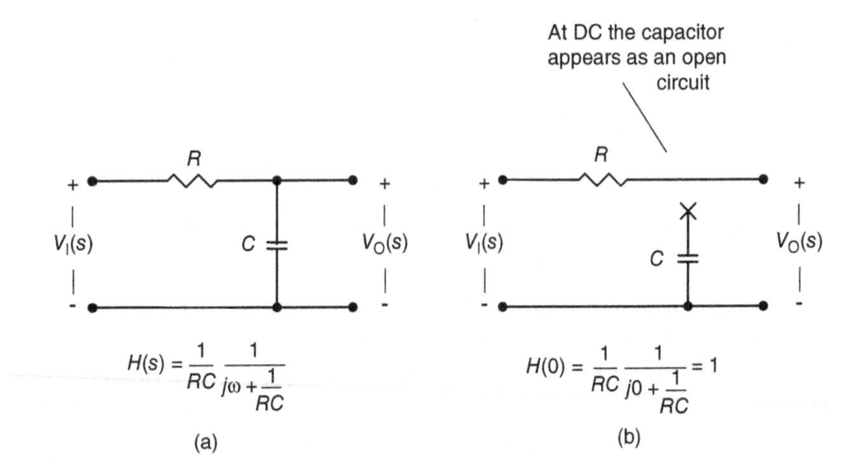

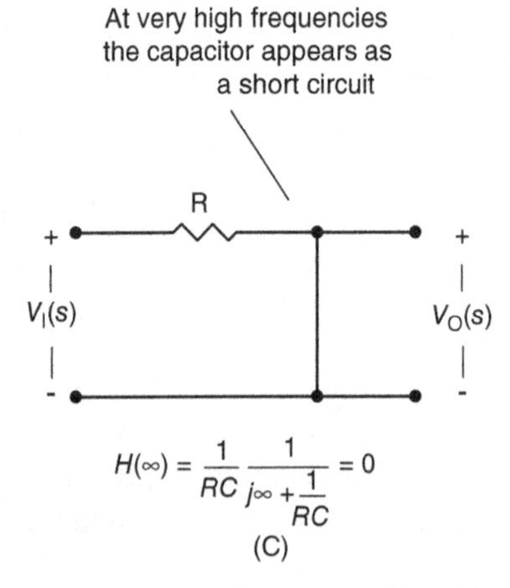

Figure 3.2. At DC, the capacitor in the lowpass RC filter (a) appears as an open circuit (b), and since no current flows through the resistor, $V_O(s)/V_I(s) = 1$. At very high frequencies, the capacitor appears as a short circuit (c) and $V_O(s)/V_I(s) = 0$.

Example 3.1. Plot the magnitude and phase of the frequency response for the general second-order transfer function

$$H(s) = \frac{\omega_n^2}{s^2 + 2\xi\omega_n s + \omega_n^2}$$

where $\xi = 0.7$ and $\omega_n = 2\pi \times 10$ MHz.

Solution. The Matlab script shown in Figure 3.3 generates the required plot. Lines 1–3 give parameters used to specify the analysis frequencies. These parameters are used later in the script to create a set of frequencies that are uniformly spaced when plotted on a logarithmic axis. Lines 6–8 allocate matrices that will hold results prior to plotting. The loop in lines 15–21 evaluates the response at each frequency point and places the results in the result matrices. Line 16 computes the analysis frequency. Line 17 computes

```
1 -   StartFreq = 1000000;    % Lowest frequency to plot.
2 -   NumDec = 3;             % Number of decades to plot.
3 -   PtsPerDec = 200;        % Number of frequency points plotted per decade.
4
5     % Pre-allocate results matricies.
6 -   MagResp = zeros(NumDec*PtsPerDec,1);     % Matrix containing magnitude response.
7 -   PhaseResp = zeros(NumDec*PtsPerDec,1);   % Matrix containing phase response.
8 -   Freq = zeros(NumDec*PtsPerDec,1);        % Matrix containing plot frequencies.
9
10    % Parameters for the general second-order transfer function to be plotted.
11 -  Wn = 2*pi*10e6;         % Natural frequency.
12 -  Zeta = 0.7;             % Damping ratio.
13
14    % Compute frequency response at frequency points uniformly-spaced on a log plot.
15 -  for i=1 : NumDec*PtsPerDec
16 -      Freq(i) = StartFreq*10^(i/PtsPerDec);   % Evaluate at this frequency.
17 -      s = 1j * 2*pi*Freq(i);                  % Determine complex frequency.
18 -      H = Wn^2/(s^2 + 2*Zeta*Wn*s + Wn^2);    % Evaluate the transfer function.
19 -      MagResp(i) = abs(H);                    % Place magnitude response in result matrix.
20 -      PhaseResp(i) = angle(H)*360/(2*pi);     % Place phase response in result matrix.
21 -  end
22
23    % Plot magnitude.
24 -  subplot(2,1,1);
25 -  HMag=semilogx(Freq,20*log10(MagResp), 'color','k', 'linewidth', 2); % Plot magnitude in dB.
26 -  set(gca,'FontSize',12);
27 -  xlabel('Freq (Hz)');
28 -  ylabel('Response (dB)');
29 -  grid on;
30
31    % Plot phase.
32 -  subplot(2,1,2);
33 -  HPhs = semilogx(Freq,PhaseResp, 'color','k', 'linewidth', 2); % Plot phase in degrees.
34 -  set(gca,'FontSize',12);
35 -  xlabel('Freq (Hz)');
36 -  ylabel('Phase (Degrees)');
37 -  ylim([-180 180]);
38 -  set(gca,'YTick',[-180, -90, 0, 90, 180]);
39 -  grid on;
```

Figure 3.3. Matlab script for plotting the frequency response of an s-domain transfer function. This technique is used throughout this chapter.

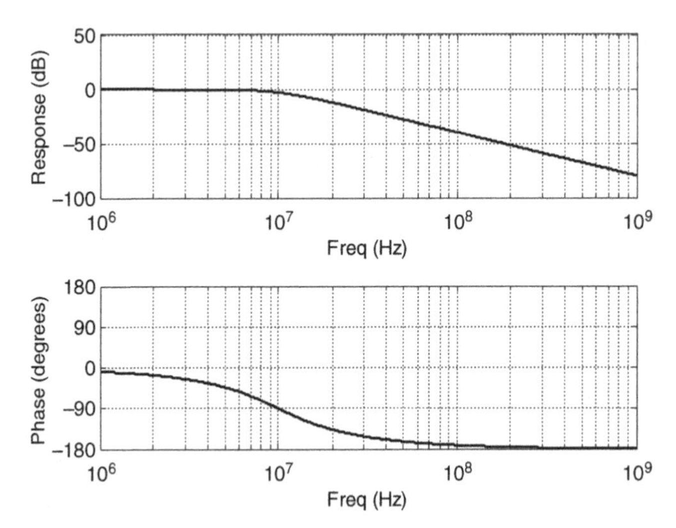

Figure 3.4. Frequency response plot showing magnitude and phase for the transfer function of Example 3.1.

the complex frequency, $s = j2\pi f$. Line 18 substitutes the complex frequency into the transfer function, and lines 19 and 20 update the result matrices. After the loop executes, lines 23–39 create the plot shown in Figure 3.4.

3.2 FREQUENCY RESPONSE AND IMPEDANCE OF SIMPLE CIRCUITS

Many simple coupling circuits can be analyzed using the steps below:

1. Reduce the circuit using superposition and/or Thevenin's theorem.
2. Write an s-domain function for the transfer function using voltage division.
3. Use Matlab to plot the frequency response directly from the s-domain function.

An advantage of this technique is that it results in a transfer function that can be analyzed using Bode analysis. Though not covered here, Bode analysis is a powerful tool for designing simple networks [1,2].

Example 3.2. The circuitry shown in Figure 3.5 is part of a phase-locked loop.[2] The error amplifier drives the circuit, and its output is fed to the

[2]This circuit is a passive second-order loop filter as described in [4].

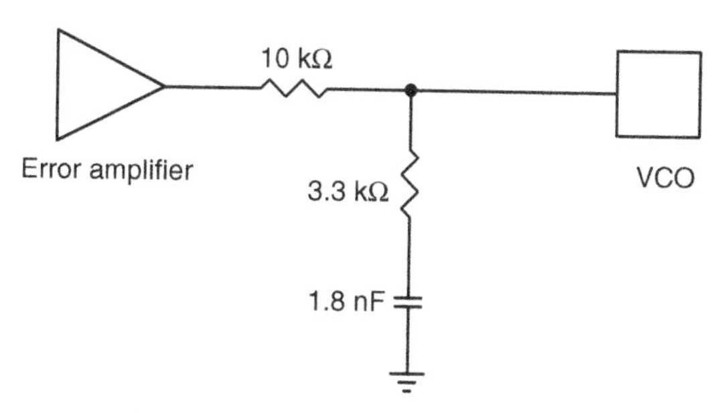

Figure 3.5. Circuit for Example 3.2.

voltage-controlled oscillator (VCO). Determine the frequency response from the error amplifier output to the VCO input.

Solution. To analyze the frequency response of this circuit, the output resistance of the error amplifier and the input resistance of the VCO must be considered. The datasheets for the parts show that they are 2.5 Ω and 500 kΩ, respectively. Figure 3.6a shows the circuit with the error amplifier represented as an ideal AC voltage source in series with 2.5 Ω and the VCO as a 500 kΩ resistance to ground. Since the series combination of 2.5 Ω and 10 kΩ is very close to 10 kΩ, the 2.5 Ω can be ignored. Similarly, since 500 kΩ is much larger than 3.3 kΩ and 10 kΩ, the 500 kΩ can also be ignored. The result of the simplifications is shown in Figure 3.6b.

The circuit in Figure 3.6b can be simplified further as seen in Figure 3.7 where R_1 is shown as Z_2 and the series combination of R_2 and C is shown as Z_1.

Using voltage division, the transfer function is written as

$$H(s) = \frac{Z_1}{Z_1 + Z_2} = \frac{R_2 + \frac{1}{sC}}{R_1 + R_2 + \frac{1}{sC}} = \frac{R_2}{R_1 + R_2} \cdot \frac{s + \frac{1}{R_2 C}}{s + \frac{1}{(R_1 + R_2)C}} \tag{3.2}$$

This equation looks somewhat complicated, but we can verify that it gives the correct response at DC and at high frequencies using asymptotic analysis. At DC the capacitor in Figure 3.6b appears as an open circuit, so inspection of the circuit of Figure 3.6b shows that the voltage gain is unity. Substituting $s = j0$ into Equation 3.2 gives

$$H(0) = \frac{R_2}{R_1 + R_2} \cdot \frac{j0 + \frac{1}{R_2 C}}{j0 + \frac{1}{(R_1 + R_2)C}} = 1 \tag{3.3}$$

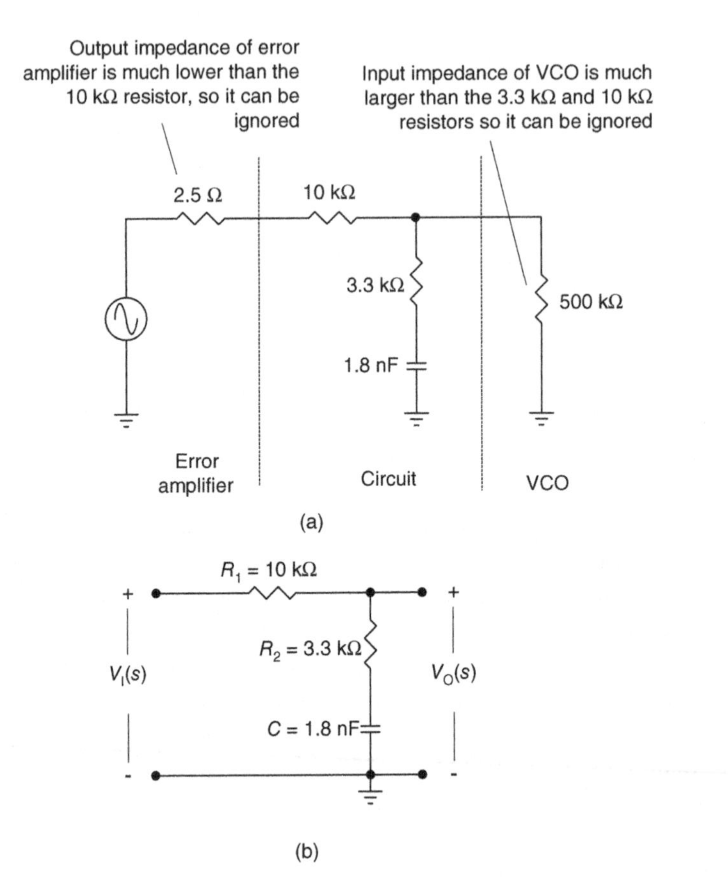

Figure 3.6. Analysis models for the circuit of Figure 3.5. The model in (a) includes the output impedance of the error amplifier and the input impedance of the VCO. These resistances are insignificant, so they are ignored and the circuit is analyzed using the model shown in (b).

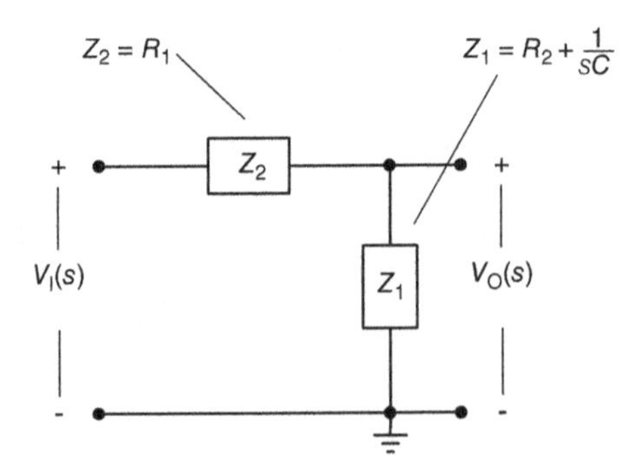

Figure 3.7. Further simplification of the circuit of Figure 3.6b. The transfer function of this circuit is obtained using voltage division.

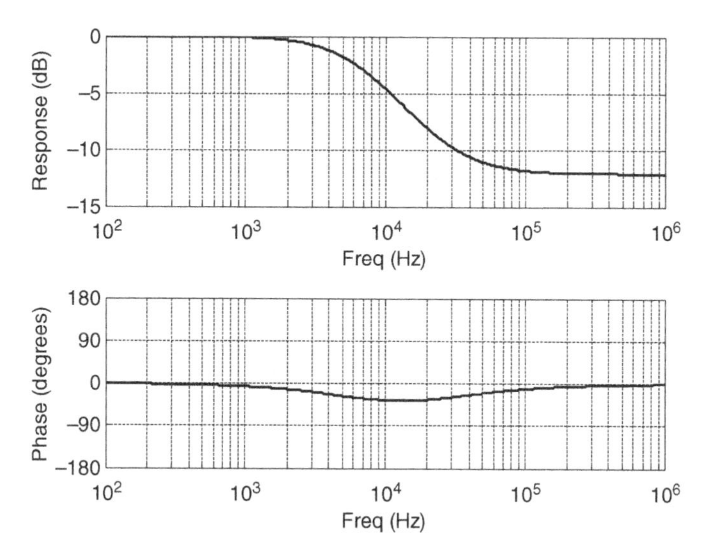

Figure 3.8. Frequency response plot showing magnitude and phase for Example 3.2.

Similarly, when the input frequency is infinite, the capacitor appears as a short circuit and inspection of Figure 3.6b gives

$$H(\infty) = \frac{R_2}{R_1 + R_2} \tag{3.4}$$

And we find that substituting $s = j\infty$ into Equation 3.2 gives the same result as Equation 3.4:

$$H(\infty) = \frac{R_2}{R_1 + R_2} \cdot \frac{j\infty + \frac{1}{R_2 C}}{j\infty + \frac{1}{(R_1+R_2)C}} = \frac{R_2}{R_1 + R_2} \tag{3.5}$$

Finally, we substitute the actual circuit values into Equation 3.2 and plot it using the Matlab code shown in Figure 3.3 resulting in the magnitude and phase plots in Figure 3.8.

Example 3.3. Determine the Thevenin voltage and impedance for the circuit of Figure 3.9a at 750 kHz. Represent the circuit as a Thevenin source as shown in Figure 3.9b.

Solution. Thevenin's theorem [2] states that a circuit consisting of linear circuit elements, such as the one in Figure 3.9a, can be represented as a voltage

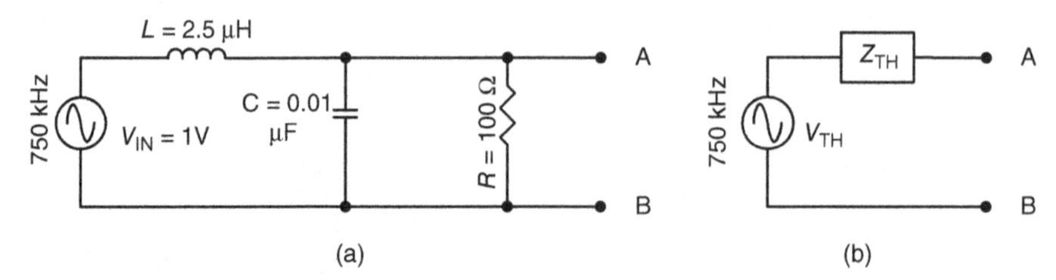

Figure 3.9. The circuit of (a) can be represented by the circuit in (b) by determining the Thevenin voltage, V_{TH}, and impedance, Z_{TH}.

source and a series impedance as shown in Figure 3.9b. Thevenin equivalents are commonly used to simplify circuits.

Determining the Thevenin voltage and impedance is done by referring to the two diagrams in Figure 3.9. If there is no external load connected to terminals A and B of Figure 3.9a and the voltage between terminals A and B is measured, the same voltage would be seen at terminals A and B in Figure 3.9b if it were similarly unloaded. Since there would be no current through Z_{TH}, this voltage would also be the Thevenin voltage V_{TH}.

If a short circuit is connected between terminals A and B in Figure 3.9a then the current would be identical to the short-circuit current of Figure 3.9b. With the Thevenin voltage in Figure 3.9b known, the Thevenin impedance can be computed as the Thevenin voltage divided by the short-circuit current from Figure 3.9a.

Our strategy to determine the Thevenin equivalent of Figure 3.9a is as follows:

1. Compute the open circuit voltage of Figure 3.9a. This is the Thevenin voltage.
2. Determine the short-circuit current of Figure 3.9a. The Thevenin impedance is the Thevenin voltage divided by this current.

To find the open circuit voltage of Figure 3.9a the transfer function must be determined. We represent the inductor impedance by Z_2 and the parallel combination of the resistor and capacitor by Z_1 as shown in Figure 3.10.

The transfer function is

$$H(s) = \frac{Z_1}{Z_1 + Z_2} = \frac{\frac{R}{1+sRC}}{\frac{R}{1+sRC} + sL} = \frac{1}{LC} \frac{1}{s^2 + \frac{s}{RC} + \frac{1}{LC}} \tag{3.6}$$

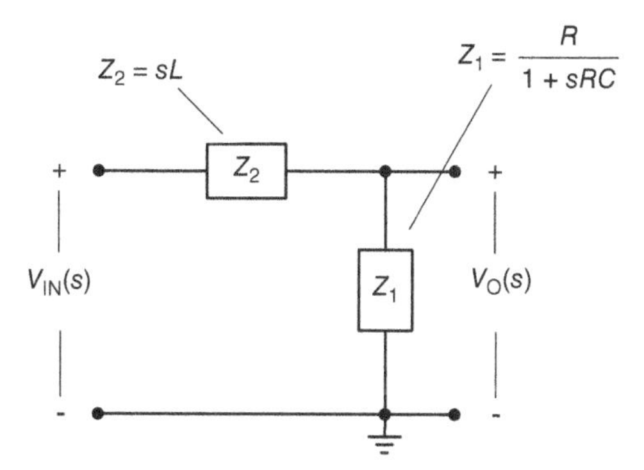

Figure 3.10. Simplification of the circuit of Figure 3.9a. The transfer function of this circuit is obtained using voltage division.

The Thevenin voltage is

$$V_{TH} = \frac{V_O(s)}{V_I(s)} \cdot V_I(s) = H(s)V_I(s) \tag{3.7}$$

The short-circuit current for the circuit of Figure 3.9a is

$$I_{SC} = \frac{V_I(s)}{sL} \tag{3.8}$$

The Thevenin impedance is the Thevenin voltage divided by the short-circuit current. This is computed using the results from Equations 3.7 and 3.8:

$$Z_{TH} = \frac{V_{TH}}{I_{SC}} = \frac{H(s)V_I(s)}{\frac{V_I(s)}{sL}} = H(s)sL \tag{3.9}$$

The lines of Matlab in Figure 3.11 are used to compute V_{TH} and Z_{TH}. The equivalent Thevenin source is shown in Figure 3.12.

Example 3.4. Someone else designed the circuit in Figure 3.13 and you just inherited it. Determine what the circuit does and if it appears to be designed correctly.

Solution. This example is poorly defined, but typical of situations you'll encounter in the workplace. Working with the designs of other engineers is an excellent way to build your skill set.

```
1    % Parameters for the circuit of Figure 9a
2    R = 100;
3    L = 2.5e-6;
4    C = 0.01e-6;
5    Vin = 1;
6
7    % Evaluate H(s)at 750 kHz.
8    s = 1j*2*pi*750e3;                      % Complex frequency
9    H = 1/(L*C)*1/(s^2+s/(R*C)+1/(L*C));   % Transfer function from Equation 6.
10
11   % Compute and display Thevenin voltage at 750 kHz.
12   VTh = H*Vin;        % Equation 7.
13
14   % Compute and display Thevenin impedance at 750 kHz.
15   ZTh = H*s*L;   % Equation 9.
```

Figure 3.11. Matlab code for computing Thevenin voltage and impedance. $V_{TH} = 2.1 - j0.56$. $Z_{TH} = 6.55 + j24.75$.

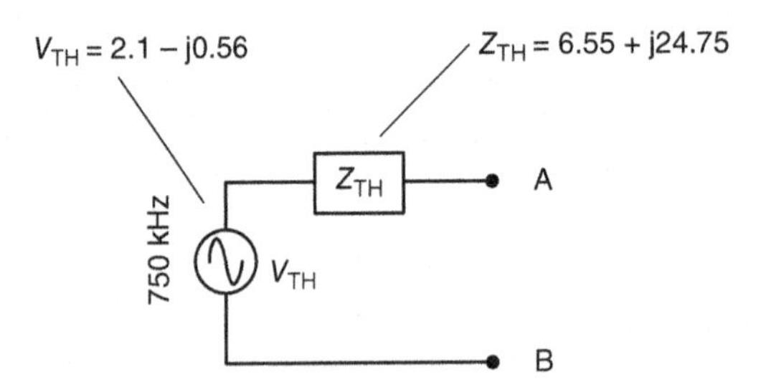

Figure 3.12. Thevenin equivalent circuit for Example 3.3 at 750 kHz.

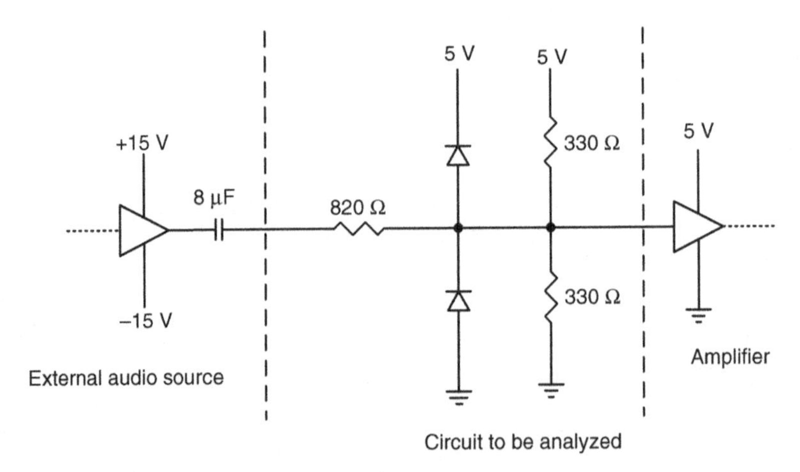

Figure 3.13. Audio coupling circuit for Example 3.4.

We begin with the clues shown on the schematic. The amplifier on the left is marked "external audio source" so we presume that the frequency of the signal of interest is in the 20 to ~15,000 Hz range. We also note that the supply voltage for the amplifier on the left is ±15 V suggesting that its output will be symmetric around 0 V. The amplifier on the right has a single positive supply voltage of +5 V meaning that its input should be centered around 2.5 V. The capacitor blocks the DC between the two amplifiers. Since the power supply for the external amplifier is ±15 V, we might expect output excursions in this range. Since the amplifier on the right has only a 5 V supply, we would expect the circuit to provide attenuation between the two amplifiers. We then note that the diodes will conduct only if the voltage at their common point lies outside of the voltage range $-0.7 \leq V \leq +5.7$. We conclude that the diodes provide protection against large voltage excursions at the input of the amplifier on the right. The 820 Ω resistor is part of the attenuator and also limits the diode current when the input voltage is out of range.

To determine if the circuit is designed correctly we need to answer the four questions below:

1. Does the circuit apply a 2.5 V DC offset at the input of the amplifier on the right?
2. Does the circuit faithfully pass audio signals in the 20 to ~15,000 Hz range?
3. Does the circuit properly attenuate the signal?
4. Does the circuit provide a sufficiently high impedance load to the external audio source?

The first three questions can be answered by computing the output of the circuit when the input is a sinewave. The analysis model is shown in Figure 3.14. Since the diodes do not affect the circuit during normal operation they are not shown.

The output is computed using the principle of superposition [2]. Our first task is to compute the output due to the 5 V supply with the signal input grounded. Next we compute the output due to a sinusoidal input with the 5 V supply grounded. The two outputs are added to get the composite circuit output.

If the sinewave input is grounded as shown in Figure 3.15a, there is only DC present and the capacitor appears as an open circuit. The resulting equivalent circuit is shown in Figure 3.15b and the output voltage due to the 5 V supply is

$$V_{O} = 5\frac{330}{330 + 330} = 2.5\text{ V} \tag{3.10}$$

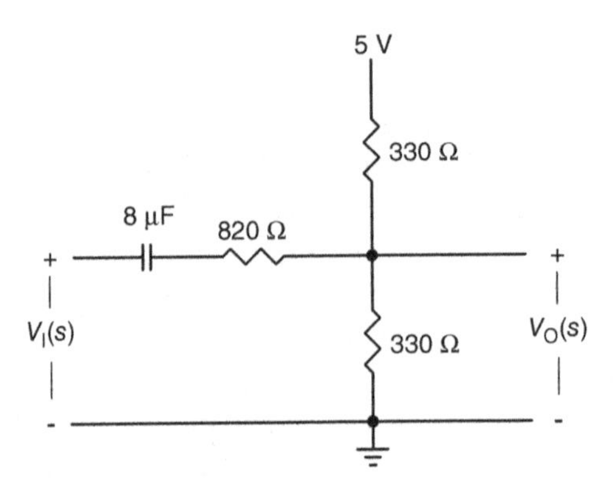

Figure 3.14. Analysis model for circuit of Figure 3.13. The diodes are not shown because they do not affect normal operation of the circuit.

This answers the first question above: The circuit indeed applies a 2.5 V DC offset at the input of the amplifier on the right.

With the 5 V supply grounded, the circuit appears as shown in Figure 3.16. The response of the circuit of Figure 3.16b to a sinewave input is

$$H(s) = \frac{R_P}{R_P + R_2 + \frac{1}{sC}} = \frac{R_P}{R_2 + R_P} \cdot \frac{s}{s + \frac{1}{(R_2+R_P)C}} \tag{3.11}$$

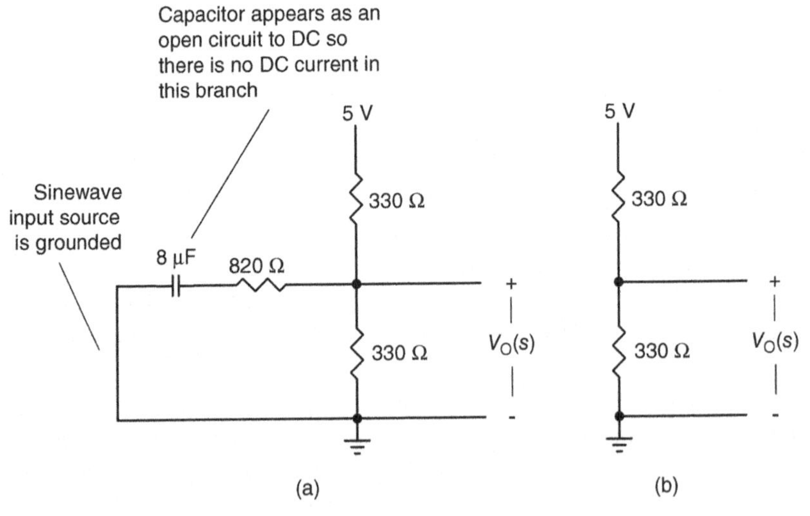

Figure 3.15. Circuit of Figure 3.14 with the sinewave input grounded (a). Since the capacitor appears as an open circuit to DC, there is no current in the leftmost branch and it is removed (b).

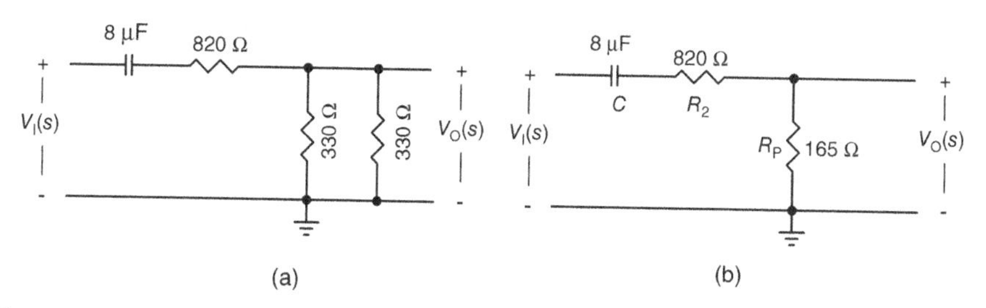

Figure 3.16. (a) Circuit of Figure 3.14 with the 5 V source grounded. (b) Circuit with 330 Ω resistors combined into R_P.

We use asymptotic analysis to check this result. At DC the capacitor in Figure 3.16b has infinite impedance so it appears as an open circuit and inspection of the circuit shows the response will be zero. Substituting $s = j0$ into Equation 3.11 validates this result. At infinite frequency, the capacitor appears as a short circuit and we expect the circuit to act like a voltage divider with gain equal to $R_P/\left(R_2 + R_P\right)$. Substituting $s = j\infty$ into Equation 3.11 validates this result as shown in Equation 3.12:

$$H(\infty) = \frac{R_P}{R_2 + R_P} \cdot \frac{j\infty}{j\infty + \frac{1}{(R_2+R_P)C}} = \frac{R_P}{R_2 + R_P} \tag{3.12}$$

Using superposition, the output voltage is equal to the result of Equation 3.10 added to the result of Equation 3.11 or

$$V_O(s) = 2.5 + H(s)V_{IN}(s) \tag{3.13}$$

The transfer function, $H(s)$, is plotted in Figure 3.17 which shows that the 3 dB highpass coupling frequency is 20 Hz and does not attenuate high frequencies. This answers our second question: The circuit does indeed pass audio signals in the 20 to ~15,000 Hz range.

Figure 3.17 shows that the circuit gain in the flat region is 0.1675 or about 1/6.[3] Therefore, if a signal from the external audio source had a voltage swing of 30 Vpp then the output signal would have a range of 5 Vpp. This answers our third question: The circuit does properly attenuate the signal.

Finally, we wish to determine if the circuit provides a sufficiently high impedance load to the amplifier on the left. We consult the manufacturer's datasheet for the amplifier and find that its output stage can drive loads greater than 2 kΩ without distorting. Referring to Figure 3.16b, and treating

[3]Note that this could also be determined from Equation 3.12.

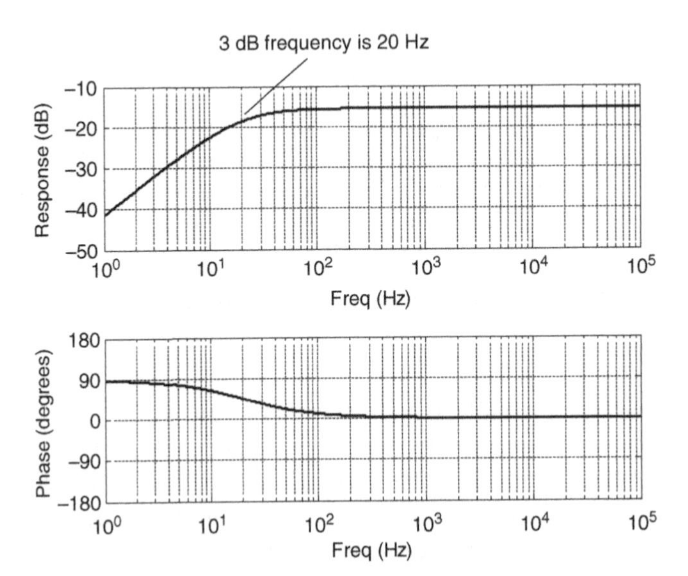

Figure 3.17. Frequency response plot showing magnitude and phase for Example 3.4.

the capacitor as a short circuit in the passband, the circuit presents a 985 Ω load to the amplifier. This answers our second question: The amplifier on the left cannot drive this network without distorting the signal.

The result of our analysis is that the circuit provides proper correct DC bias, attenuation, and frequency response. However, it presents an excessive load to the amplifier that drives it and therefore does not appear to be designed correctly.

Example 3.5. The circuit shown in Figure 3.18 is a loop filter for a switching power supply. Determine the voltage transfer function $V_O(s)/V_{IN}(s)$ and plot the frequency response.

Solution. The amplifier in Figure 3.18 is a transconductance amplifier meaning that its output current is equal to the product of the transconductance, g_m, and input voltage, V_1. Figure 3.19 shows that the amplifier isolates the circuitry on the left from the circuitry on the right thereby simplifying the analysis.

Referring to Figure 3.19, the transfer function for the circuit can be determined as shown below:

$$V_1(s) = V_{IN}(s)H_1(s) \tag{3.14}$$

and

$$V_O(s) = g_m V_1(s)Z_2(s) \tag{3.15}$$

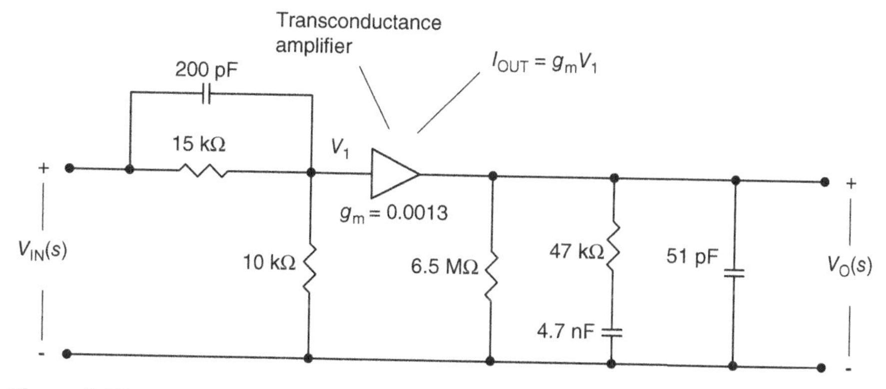

Figure 3.18. Loop filter circuit for Example 3.5. The transconductance amplifier simplifies analysis because it isolates the circuitry on the left from the circuitry on the right.

Substituting Equation 3.14 into Equation 3.15 and solving for the transfer function gives

$$H(s) = \frac{V_O(s)}{V_{IN}(s)} = g_m H_1(s) Z_2(s) \tag{3.16}$$

Our strategy for determining the transfer function is to compute $H_1(s)$ and $Z_2(s)$, and then substitute them into Equation 3.16. Figure 3.20 shows the circuit redrawn to show reference designators used for analysis.

To determine $H_1(s)$, we begin by computing the impedance of the parallel combination of C_A and R_{2A}:

$$Z_P = \frac{R_{2A}\frac{1}{sC_A}}{R_{2A} + \frac{1}{sC_A}} = \frac{R_{2A}}{1 + sR_{2A}C_A} \tag{3.17}$$

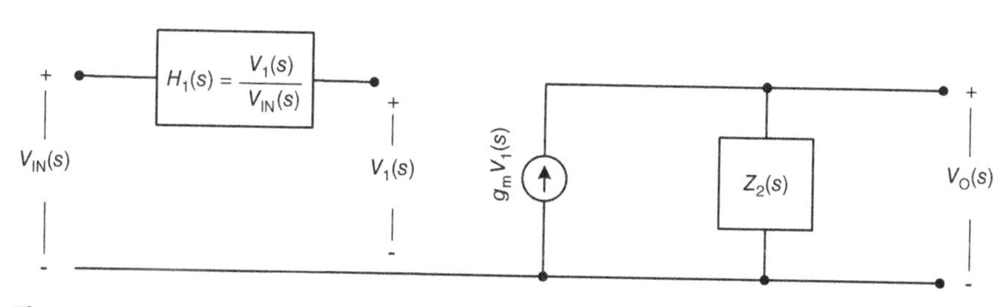

Figure 3.19. Circuit of Figure 3.18 redrawn to show the circuit model of the transconductance amplifier and the isolated transfer functions.

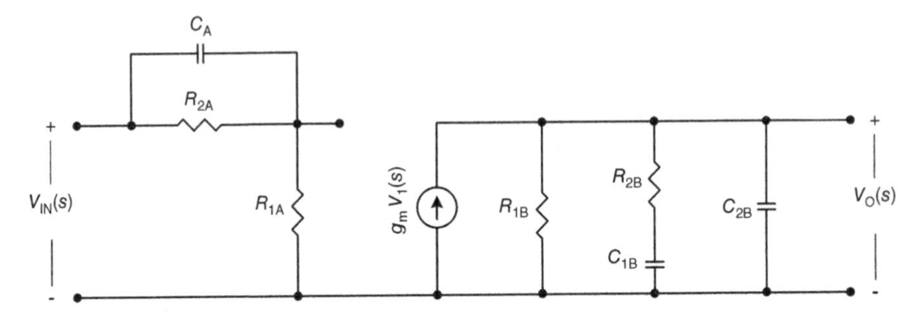

Figure 3.20. Circuit of Figure 3.18 redrawn to show reference designators. The transconductance amplifier is represented by its circuit model.

Then use voltage division to get $H_1(s)$:

$$H_1(s) = \frac{R_{1A}}{R_{1A} + Z_P} = \frac{R_{1A}}{R_{1A} + R_{2A}} \cdot \frac{1 + sC_A R_{2A}}{1 + \frac{sC_A R_{1A} R_{2A}}{R_{1A} + R_{2A}}} \tag{3.18}$$

We use asymptotic analysis to check Equation 3.18 at DC and infinite frequency. Referring to Figure 3.20 it is seen that at DC, capacitor C_{2A} appears as an open circuit, and the transfer function can be determined using voltage division and resistors R_{1A} and R_{2A}. At very high frequencies, capacitor C_{2A} appears as a short circuit and the transfer function is unity. The reader is encouraged to verify this from Equation 3.18.

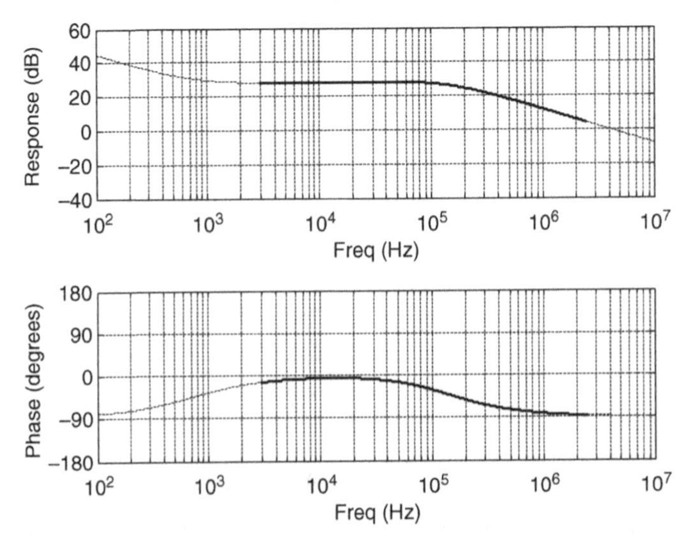

Figure 3.21. Frequency response plot showing magnitude and phase for Example 3.5.

To determine $Z_2(s)$ recall that the parallel combination of elements Z_A, $Z_B, \ldots, Z_N$ is

$$Z_{\text{Parallel}} = \frac{1}{\frac{1}{Z_A} + \frac{1}{Z_B} + \cdots + \frac{1}{Z_N}} \tag{3.19}$$

Using Equation 3.19 and Figure 3.20, the expression for $Z_2(s)$ is written by inspection:

$$Z_2 = \frac{1}{\frac{1}{R_{1B}} + \frac{1}{R_{2B}+(1/sC_{1B})} + sC_{2B}} \tag{3.20}$$

Substituting Equations 3.18 and 3.20 into Equation 3.16 gives the final transfer function:

$$H(s) = \frac{R_{1A}}{R_{1A} + R_{2A}} \cdot \frac{1 + sC_A R_{2A}}{1 + \frac{sC_A R_{1A} R_{2A}}{R_{1A}+R_{2A}}} g_m \cdot \frac{1}{\frac{1}{R_{1B}} + \frac{1}{R_{2B}+(1/sC_{1B})} + sC_{2B}} \tag{3.21}$$

The frequency response for the circuit is plotted in Figure 3.21.

3.3 FREQUENCY RESPONSE FOR LADDER NETWORKS

The analysis techniques presented in Section 3.2 will enable you to quickly and accurately analyze most circuits you'll encounter as a working engineer. However, you will also encounter circuits of greater complexity, and the techniques of the previous section will not suffice. This section presents a method for analyzing circuits consisting of alternating series and shunt branches shown in Figure 3.22 and called *ladder networks*. The technique

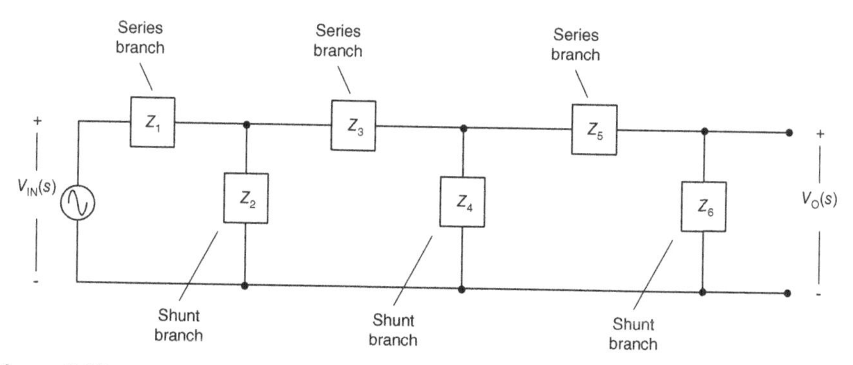

Figure 3.22. Ladder network topology. Ladder networks of any length can be quickly analyzed using the method shown in this section.

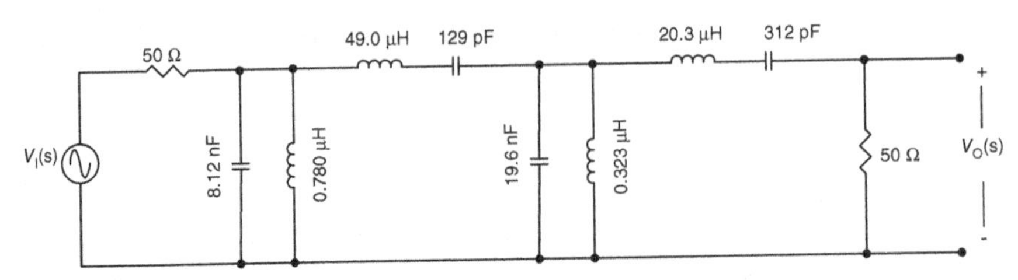

Figure 3.23. Ladder circuit. The structure of this circuit simplifies analysis.

shown here will enable you to analyze most passive filters which are typically implemented as ladder networks.

Example 3.6. Determine the frequency response of the circuit shown in Figure 3.23.

Solution. The analysis model is shown in Figure 3.24. Our strategy is to first specify a voltage of 1 V for $V_O(s)$ and then work backwards to determine $V_{IN}(s)$.[4] This is done using the procedure shown in the equations below at each frequency. The dependence on complex frequency, s, is omitted for clarity:

$$i_1 = \frac{V_O}{50} = \frac{1}{50} \tag{3.22}$$

$$V_A = V_O + i_1 Z_4 \tag{3.23}$$

$$i_2 = \frac{V_A}{Z_3} \tag{3.24}$$

$$i_3 = i_1 + i_2 \tag{3.25}$$

$$V_B = V_A + i_3 Z_2 \tag{3.26}$$

$$i_4 = \frac{V_B}{Z_1} \tag{3.27}$$

$$i_5 = i_4 + i_3 \tag{3.28}$$

$$V_{IN} = V_B + i_5 50 \tag{3.29}$$

The transfer function is

$$H(s) = \frac{V_O}{V_{IN}} = \frac{1}{V_{IN}} \tag{3.30}$$

The lines of Matlab used to evaluate the transfer function are shown in Figure 3.25. The frequency response is plotted in Figure 3.26 using the lines of Matlab code at the bottom of Figure 3.3.

[4]Therefore the transfer function $H(s) = 1/V_{IN}(s)$.

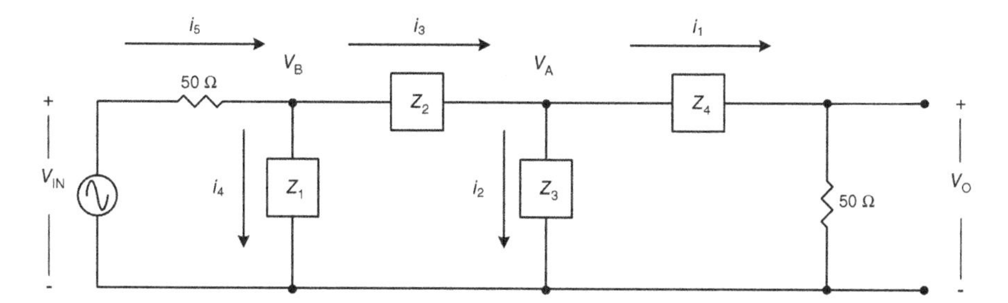

Figure 3.24. Analysis model for the ladder circuit shown in Figure 3.23.

```
1    % Specify components.
2    Rs = 50;                                    % Source resistance.
3    C1 = 8.12e-9;    L1 = 0.780e-6;             % Impedance Z1
4    C2 = 129e-12;    L2 = 49.0e-6;              % Impedance Z2.
5    C3 = 19.6e-9;    L3 = 0.323e-6;             % Impedance Z3.
6    C4 = 312e-12;    L4 = 20.3e-6;              % Impedance Z4.
7    RL = 50;                                    % Termination resistance.
8
9
10   StartFreq = 0.1e6;  % Lowest frequency to plot.
11   NumDec = 2;         % Number of decades to plot.
12   PtsPerDec = 200;    % Number of frequency points plotted per decade.
13
14   % Pre-allocate results matricies.
15   MagResp = zeros(NumDec*PtsPerDec,1);     % Matrix containing magnitude response.
16   PhaseResp = zeros(NumDec*PtsPerDec,1);   % Matrix containing phase response.
17   Freq = zeros(NumDec*PtsPerDec,1);        % Matrix containing plot frequencies.
18
19
20   % Compute frequency response at frequency points uniformly-spaced on a log plot.
21   for i=1 : NumDec*PtsPerDec
22       Freq(i) = StartFreq*10^(i/PtsPerDec);    % Evaluate at this frequency.
23       s = 1j * 2*pi*Freq(i);                   % Determine complex frequency.
24
25       % Compute branch impedances at this frequency.
26       Z1 = s*L1/(s^2*L1*C1+1);
27       Z2 = s*L2 + 1/(s*C2);
28       Z3 = s*L3/(s^2*L3*C3+1);
29       Z4 = s*L4 + 1/(s*C4);
30
31       i1 = 1/RL;            % Equation 22
32       Va = 1 + i1*Z4;       % Equation 23
33       i2 = Va/Z3;           % Equation 24
34       i3 = i2 + i1;         % Equation 25
35       Vb = Va + i3*Z2;      % Equation 26
36       i4 = Vb/Z1;           % Equation 27
37       i5 = i3 + i4;         % Equation 28
38       Vin = Vb + i5*Rs;     % Equation 29
39       H = 1/Vin;            % Equation 30
40
41       MagResp(i) = abs(H);                      % Place magnitude response in result matrix.
42       PhaseResp(i) = angle(H)*360/(2*pi);       % Place phase response in result matrix.
43   end
```

Figure 3.25. Matlab code for plotting the frequency response of the ladder network of Figure 3.23.

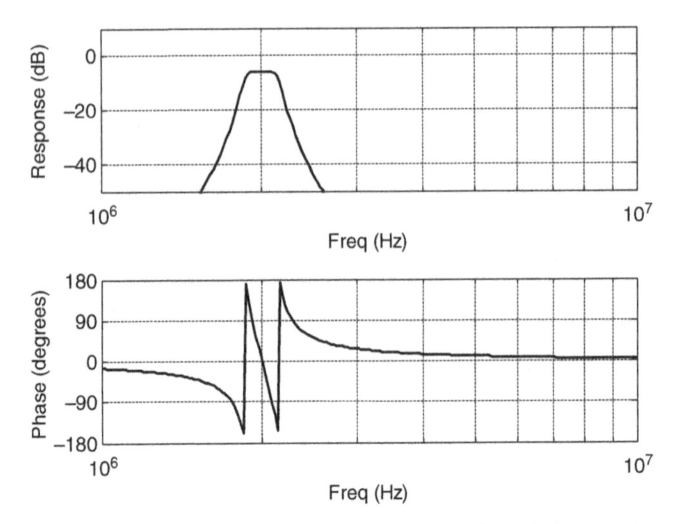

Figure 3.26. Frequency response plot showing magnitude and phase for Example 3.6.

3.4 GENERALIZED TECHNIQUE FOR DETERMINING FREQUENCY RESPONSE

This section presents a generalized method for analyzing circuits, where the fundamentals of circuit analysis are used to generate equations and a computational tool such as Matlab is used for the computations. The basic strategy is to use nodal or mesh analysis to create a set of simultaneous equations where the node voltages or mesh currents are the unknown variables, and the impedances of the components and the voltages or currents of the signal generators are known. The set of equations is solved at each frequency point and the resulting node voltages or mesh currents are used to compute the desired result.

Example 3.7. Determine the frequency response of the circuit shown in Figure 3.27.[5]

Solution. The circuit is first redrawn in Figure 3.28 to show generalized impedances and mesh currents I_1, I_2, and I_3.

The voltage drop across any component is equal to the sum of the mesh currents going through it multiplied by the impedance of the component. Using Kirchhoff's law, the sum of the voltage drops in each closed path is zero.[6]

[5]This clever notch filter circuit can be found on page 6-19 of [3]. The 680 Ω resistor significantly increases the notch depth.

[6]This is the same method used for solving resistive circuits in Chapter 1.

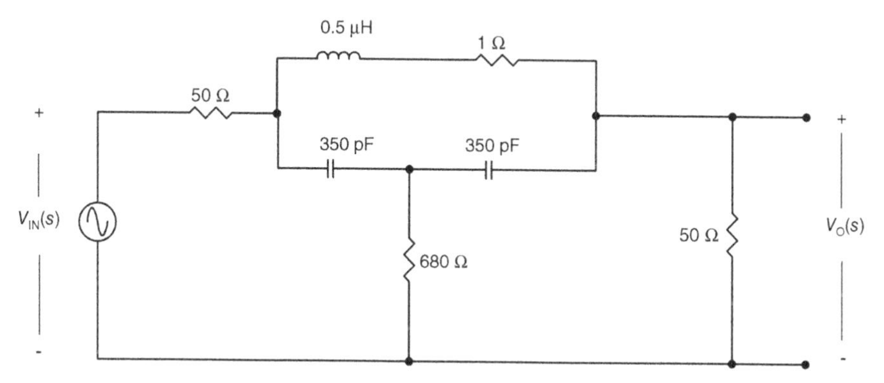

Figure 3.27. Circuit for Example 3.7.

The equation for the leftmost loop is

$$-V_{IN} + I_1Z_1 + I_1Z_2 - I_2Z_2 + I_1Z_3 - I_3Z_3 = 0 \tag{3.31}$$

Or

$$(Z_1 + Z_2 + Z_3)I_1 + (-Z_2)I_2 + (-Z_3)I_3 = V_{IN} \tag{3.32}$$

The equation for the upper loop is

$$I_2Z_2 - I_1Z_2 + I_2Z_4 + I_2Z_5 + I_2Z_6 - I_3Z_6 = 0 \tag{3.33}$$

Or

$$(-Z_2)I_1 + (Z_2 + Z_4 + Z_5 + Z_6)I_2 + (-Z_6)I_3 = 0 \tag{3.34}$$

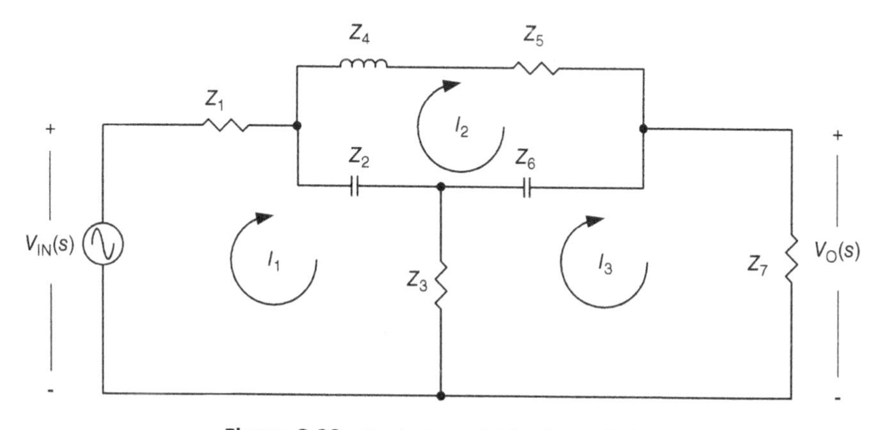

Figure 3.28. Analysis model for Example 3.7.

The equation for the rightmost loop is

$$I_3Z_3 - I_1Z_3 + I_3Z_6 - I_2Z_6 + I_3Z_7 = 0 \tag{3.35}$$

or

$$(-Z_3)\,I_1 + (-Z_6)\,I_2 + (Z_3 + Z_6 + Z_7)\,I_3 = 0 \tag{3.36}$$

Equations 3.32, 3.34, and 3.36 can be combined into a single matrix equation in the form

$$[Z]\,[I] = [V] \tag{3.37}$$

From Equations 3.32, 3.34, and 3.36, matrices $[\mathbf{Z}]$, $[\boldsymbol{I}]$, and $[\boldsymbol{V}]$ are

$$[\mathbf{Z}] = \begin{bmatrix} Z_1 + Z_2 + Z_3 & -Z_2 & -Z_3 \\ -Z_2 & Z_2 + Z_4 + Z_5 + Z_6 & -Z_6 \\ -Z_3 & -Z_6 & Z_3 + Z_6 + Z_7 \end{bmatrix} \tag{3.38}$$

$$[\boldsymbol{I}] = \begin{bmatrix} I_1 \\ I_2 \\ I_3 \end{bmatrix} \tag{3.39}$$

$$[\boldsymbol{V}] = \begin{bmatrix} V_{\text{IN}} \\ 0 \\ 0 \end{bmatrix} \tag{3.40}$$

Equation 3.37 can be solved for$[\boldsymbol{I}]$ by premultiplying both sides by $[\mathbf{Z}]^{-1}$ which gives

$$[\mathbf{Z}]^{-1}[\mathbf{Z}][\boldsymbol{I}] = [\mathbf{Z}]^{-1}[\boldsymbol{V}] \tag{3.41}$$

or

$$[\boldsymbol{I}] = [\mathbf{Z}]^{-1}[\boldsymbol{V}] \tag{3.42}$$

The Matlab code shown in Figure 3.29 creates matrix $[\mathbf{Z}]$ at each frequency and solves the matrix. Since the input voltage is set to 1 in line 36, the transfer function is equal to the output voltage. The frequency response is shown in Figure 3.30 and plotted using the lines of Matlab code at the bottom of Figure 3.3.

```
1       % Component values
2       R1 = 50;
3       C2 = 350e-12;
4       R3 = 680;
5       L4 = 0.5e-6;
6       R5 = 1;
7       C6 = 350e-12;
8       R7 = 50;
9
10      StartFreq = 1e6;     % Lowest frequency to plot.
11      NumDec = 2;          % Number of decades to plot.
12      PtsPerDec = 200;     % Number of frequency points plotted per decade.
13
14      % Pre-allocate results matricies.
15      MagResp = zeros(NumDec*PtsPerDec,1);      % Matrix containing magnitude response.
16      PhaseResp = zeros(NumDec*PtsPerDec,1);    % Matrix containing phase response.
17      Freq = zeros(NumDec*PtsPerDec,1);         % Matrix containing plot frequencies.
18
19      for i = 1 : NumDec*PtsPerDec
20          Freq(i) =  StartFreq*10.^(i/PtsPerDec);      % Evaluate at this frequency.
21          s = 1j*2*pi*Freq(i);                         % Determine complex frequency.
22
23          % Compute impedances of each component at this frequency.
24          Z1 = R1;
25          Z2 = 1/(s*C2);
26          Z3 = R3;
27          Z4 = s*L4;
28          Z5 = RL;
29          Z6 = 1/(s*C6);
30          Z7 = R7;
31
32          Z = [Z1+Z2+Z3,  -Z2,           -Z3; ...      % Create matrix Z
33                -Z2,       Z2+Z4+Z5+Z6,  -Z6; ...
34                -Z3,       -Z6,          Z3+Z6+Z7];
35
36          V = [1; 0; 0];       % Create matrix V. Input voltage is 1 V.
37          I = Z^-1*V;          % I contains the three mesh currents at this frequency.
38
39          % Since the input is 1V, the transfer function is just I3 multiplied by R7.
40          MagResp(i) = abs(I(3)*Z7);                     % Place magnitude response in result matrix.
41          PhaseResp(i) = angle(I(3)*Z7)*360/(2*pi);      % Place phase response in result matrix.
42      end
```

Figure 3.29. Matlab code to compute frequency response for Example 3.7.

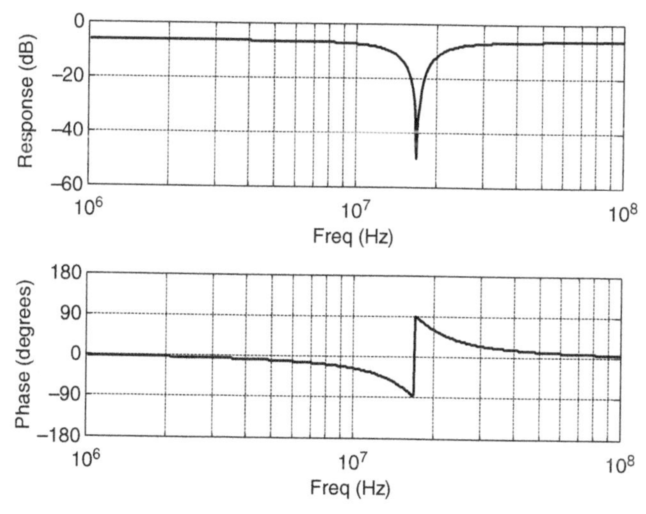

Figure 3.30. Frequency response plot showing magnitude and phase for Example 3.7.

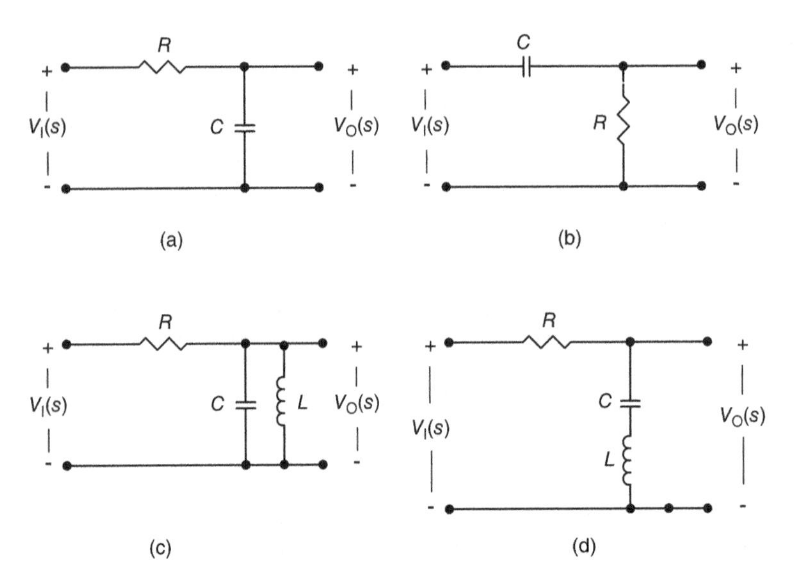

Figure 3.31. Circuits for Problem 3.1.

PROBLEMS

3.1 For each circuit (Figure 3.31)

(a) Determine the transfer function $V_O(s)/V_{IN}(s)$.

(b) Check your transfer function using asymptotic analysis.

(c) Sketch the frequency response.

3.2 Determine and plot the transfer function $V_O(s)/V_{IN}(s)$ for the circuit in Figure 3.32 using one of the following strategies:

(a) Create a Thevenin model of R_2, C, and the input source. Then connect R_1 and L, and use voltage division.

(b) Determine the parallel combination of C and the series combination of R_1 and L. Then use voltage division.

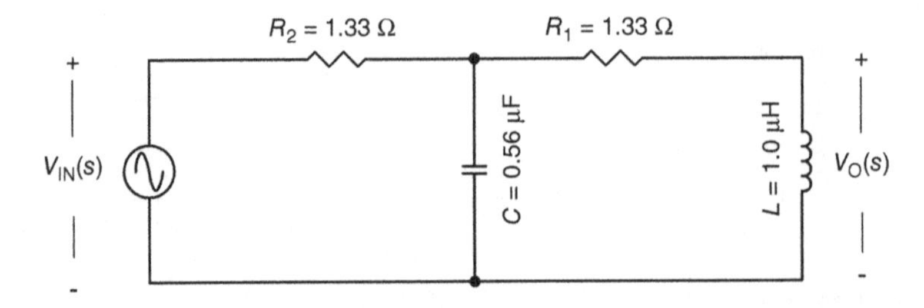

Figure 3.32. Circuit for Problem 3.2.

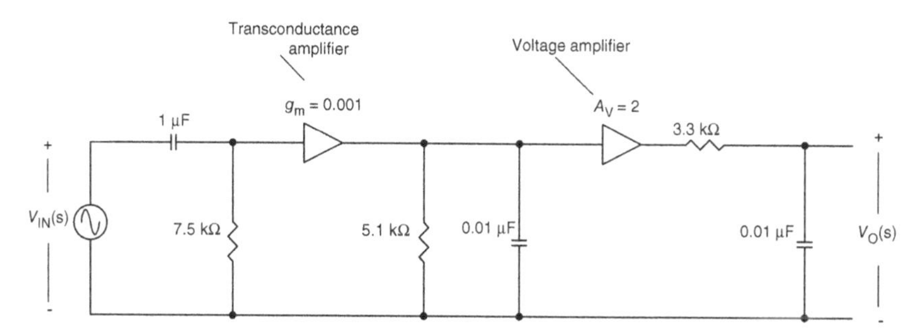

Figure 3.33. Circuit for Problem 3.4.

3.3 The high-pass coupling network in Example 3.4 was found to be unsuitable because its input impedance was too low for the amplifier driving it. Update the circuit so that its input impedance is 5 kΩ and all other characteristics are unchanged. *Hint:* Start with Figure 3.16b or see Section 1.2 for design equations for this network.

3.4 Determine the transfer function for the audio circuit shown in Figure 3.33. *Hint:* Note that the amplifiers decouple the circuits thereby simplifying analysis.

3.5 In Example 3.5 we determined and plotted the transfer function for a loop filter circuit. However, our transfer function was not in a form that provided insight about the circuit. Show that Equation 3.21, the transfer for the network, can be expressed as

$$H(s) = \frac{1}{C_{2B}} \frac{s + \frac{1}{\tau_1}}{s + \frac{1}{\tau_2}} g_m \frac{s + \frac{1}{\tau_3}}{s^2 + 2\xi\omega_n s + \omega_n^2} \tag{3.43}$$

where

$$\tau_1 = R_{2A} C_A \tag{3.44}$$

$$\tau_2 = \frac{R_{1A} R_{2A}}{R_{1A} + R_{2A}} C_A \tag{3.45}$$

$$\tau_3 = R_{2B} C_{1B} \tag{3.46}$$

$$\omega_n = \frac{1}{\sqrt{R_{1B} R_{2B} C_{1B} C_{2B}}} \tag{3.47}$$

$$\xi = \frac{R_{2B} C_{1B} + R_{1B} C_{1B} + R_{1B} C_{2B}}{2\sqrt{R_{1B} R_{2B} C_{1B} C_{2B}}} \tag{3.48}$$

3.6 Plot the frequency response for the circuit in Problem 3.2 using the ladder method.

3.7 Plot the frequency response for the circuit in Problem 3.2 using mesh currents.

3.8 Plot the input impedance for the circuit of Example 3.7. (*Hint:* Refer to Figure 3.28.) The input impedance is $Z_{IN}(s) = V_{IN}(s)/I_1(s)$. Check your plot at high and low frequencies by inspection of the circuit at DC and infinite frequency.

REFERENCES

1. Wikipedia. *Bode Plot*. Available: http://en.wikipedia.org/wiki/Bode_Plot (Accessed 11-30-2013.)
2. C. K. Alexander, *Fundamentals of Electric Circuits*, McGraw Hill, 2012.
3. A. B. Williams, *Electronic Filter Design Handbook*, McGraw-Hill, 1981.
4. F. Gardner, *Phaselock Techniques*, John Wiley & Sons, 1979.

4

HOW TO USE STATISTICS TO ENSURE A MANUFACTURABLE DESIGN

A *manufacturable* product is one that can be easily reproduced hundreds or thousands of times on the manufacturing floor and, when placed in service, perform properly and not be excessively prone to failure. Design for manufacturability (DFM) is a valuable skill because it prevents expensive and time-consuming design changes, production line changes, warranty expenses, and damage to the company's reputation.

After a product is designed, it is common to build a small number of prototype units. These units are used for debugging and testing, but since they represent only a small statistical sample, they are not sufficient for establishing whether the product is actually manufacturable. Fortunately component manufacturer's datasheets include information about the variability of their parts so that designers can estimate failure probabilities when hundreds or thousands of the parts are used. This chapter will help you understand the manufacturer's specifications and show how to estimate failure probabilities of systems consisting of multiple parts. These techniques will help you design

Ten Essential Skills for Electrical Engineers, First Edition. Barry L. Dorr.

products that will succeed both on the manufacturing floor and ultimately in the hands of your customers.

Technical interviews usually cover basic statistics because most companies have had experiences with products that failed because they could not be reliably manufactured. Making design changes once a product is on the manufacturing floor is extremely expensive, so companies want to insure that new hires can use statistics during the design phase.

This chapter reviews several basic skills related to manufacturability. We begin by using the concept of statistical independence to determine the probability of a system failure due to failure of a single component. We then review how to use the Gaussian or normal distribution and show how it can be used to solve practical manufacturing problems such as establishing test limits. Finally we show how statistics can be used to specify and design a custom component.

4.1 INDEPENDENT COMPONENT FAILURES

In this section we solve a practical problem using the concept of *statistical independence*.

If events are statistically independent then the probability of all events occurring is the product of the probabilities of the individual events occurring. Loosely speaking, events are statistically independent if the outcome of one event has no effect on the probability of any of the other events. For example, say we have a bin of transistors and we are told the probability of picking a bad one is 0.01 or 1%. Assuming the bin is well mixed, meaning bad transistors are fully interspersed with good ones, the probability of picking a good one is not affected by whether any other one picked was good or bad. The probability of picking two bad ones in a row is $0.01 \times 0.01 = 0.0001$ or 0.01%, the probability of picking three in a row is $0.01 \times 0.01 \times 0.01 = 0.0001\%$, and so forth.

This concept is important to our discussion because frequently systems require all (or nearly all) of their components to work for the system to work. If the probability of each constituent device not working after it is stuffed onto the circuit board is known, and if device failures are independent, then the manufacturing reject probability of the assembly can be estimated as shown in Example 4.1.

Example 4.1. An antenna tuner circuit board constructs RF matching networks[1] by switching inductors and capacitors in and out of the Pi network

[1]Matching networks are discussed in Chapter 9.

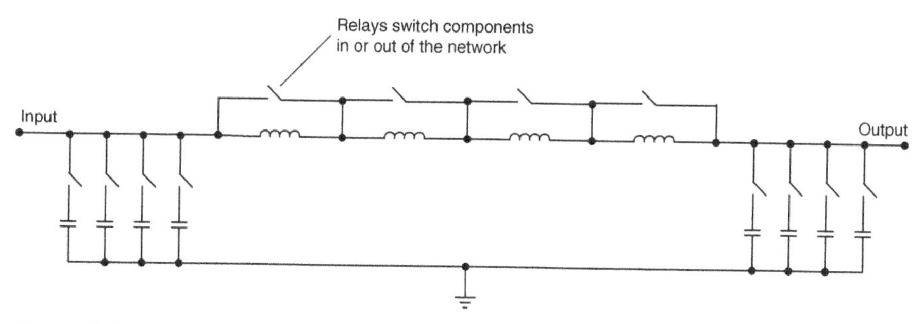

Figure 4.1. Antenna tuner Pi network. All relays must work properly for the board to pass the manufacturing test.

with relays as shown in Figure 4.1. The manufacturing test for the tuner requires every relay to work properly. It is known that after insertion in the PC board the probability of a relay not working is 0.1% or 0.001. What is the probability of a board failing manufacturing test due to a defective relay?

Solution. The probability of a defective relay is 0.001 so the probability that a relay is good is

$$P_G = 1 - P_F = 1 - 0.001 = 0.999 \tag{4.1}$$

A board will pass the manufacturing test if and only if all the relays are good. There are 12 relays on the board and we assume defects are independent so

$$P\,(\text{all relays good}) = P_G^{12} = 0.999^{12} = 0.9881 \tag{4.2}$$

The probability of the board failing the manufacturing test is

$$\begin{aligned} P\,(\text{fail manufacturing test}) &= 1 - P\,(\text{all relays good}) \\ &= 1 - 0.9981 = 0.0119 = 1.2\% \end{aligned} \tag{4.3}$$

4.2 USING THE GAUSSIAN DISTRIBUTION

Many processes in electronic manufacturing follow the normal or Gaussian distribution.[2] For example, if the voltage at the same test point on 100 different circuit boards is measured, the data will very likely be Gaussian distributed. Alternatively, the measurement could be the 3 dB bandwidth of a filter, the

[2]The terms "normal" and "Gaussian" are used interchangeably in this chapter.

value of an inductor, and so on, and the results would again follow a Gaussian distribution.[3]

The Gaussian distribution is a powerful tool because:

1. The statistical behavior of a Gaussian-distributed process is fully defined by just two parameters, mean, m, and standard deviation, σ. If these parameters are known, other useful information about the process can be extracted.
2. If samples from different Gaussian-distributed processes are added together, the result is another Gaussian-distributed process.

Using the Gaussian distribution we can

1. Combine multiple Gaussian-distributed error sources to determine expected variations of a measurement at a circuit test point.
2. Measure data, extract the mean and standard deviation, and then use the Gaussian distribution to make meaningful predictions about the process.

The probability density function (or pdf) for the Gaussian distribution is shown in Figure 4.2. The equation for the curve is

$$f(x, m, \sigma) = \frac{1}{\sigma\sqrt{2\pi}} e^{-1/2(x-m/\sigma)^2} \tag{4.4}$$

where m is the mean or expected value and σ is the standard deviation.

The cumulative distribution function (cdf) is obtained by integrating the Gaussian pdf in Figure 4.3a from $-\infty$ to c and gives the probability that the random sample is less than c. The cumulative distribution function, or cdf, is shown in Figure 4.3b.

There is no closed form solution of the integral of the pdf in Equation 4.4, so lookup tables must be used to compute cdfs. Tables are tabulated with the mean set equal to zero. The most common table tabulates the error (or erf) function[4]:

$$\mathrm{erf}\left(\frac{k}{\sqrt{2}}\right) = \text{Probability}(x \text{ lies with in } \pm\, k\sigma) \tag{4.5}$$

The example below shows how to solve a practical problem using erf( ).

[3]This is a consequence of the *central limit theorem* which provides insight into why the Gaussian distribution describes so many processes in nature.

[4]Erf( ) is a built-in function for both Matlab and Excel.

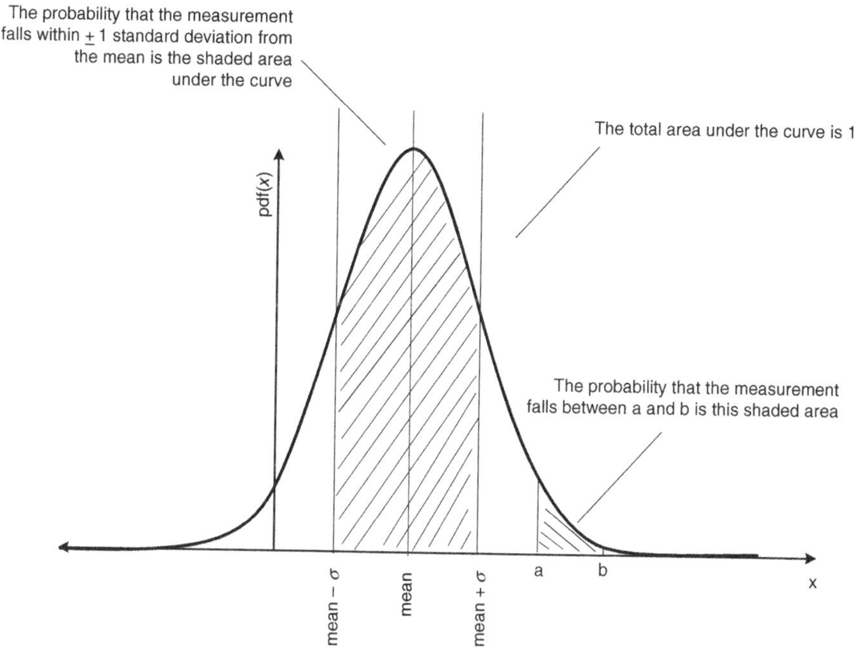

Figure 4.2. Gaussian probability density function (pdf). The area under the curve between two points on the horizontal axis is the probability that the measurement falls between the two points.

Example 4.2. The voltage at a circuit test point is known to be Gaussian distributed with mean value 1 V and $\sigma = 30$ mV. What is the probability of observing a value less than 0.96 V?

Solution. 0.96 V is 40 mV below the mean. Expressed in terms of σ it is

$$-40\text{ mV} = -\frac{40\text{ mV}}{\sigma}\cdot\sigma = -\frac{40\text{ mV}}{30\text{ mV}}\cdot\sigma = -1.3\bar{3}\sigma \tag{4.6}$$

Using Equation 4.5 we get the probability that the measurement would be within the range $\pm 1.3\bar{3}\sigma$ as shown in Figure 4.4a:

$$P\left(-1.3\bar{3}\sigma \le V \le -1.3\bar{3}\sigma\right) = \text{erf}\left(\frac{1.3\bar{3}}{\sqrt{2}}\right) = 0.8176 \tag{4.7}$$

The probability that the voltage is not within the range of $\pm 1.3\bar{3}\sigma$ is shown in Figure 4.4b and computed as

$$1 - P\left(-1.3\bar{3}\sigma \le V \le -1.3\bar{3}\sigma\right) = 1 - 0.8176 = 0.1824 \tag{4.8}$$

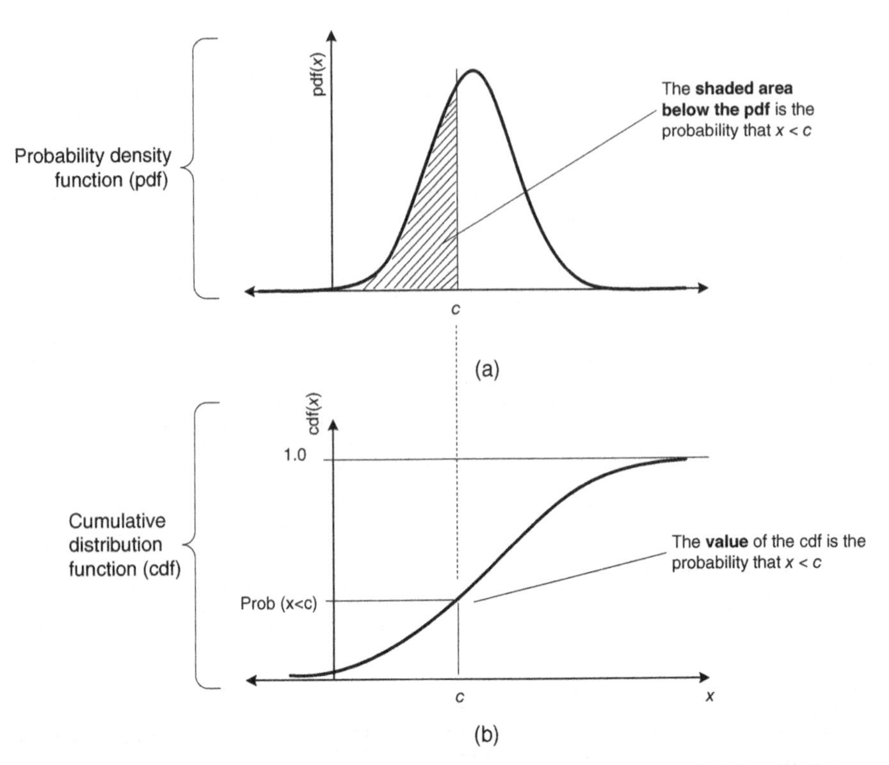

Figure 4.3. Integrating the Gaussian pdf (a) from $-\infty$ to c gives the probability that a random sample will fall below c. The cumulative distribution (b) shows the probability that a random sample will fall below the value on the horizontal axis.

Since the distribution is symmetrical, we get the final answer by dividing the shaded area in Figure 4.4b by 2 which results in Figure 4.4c. The final answer is

$$P(V \leq 0.96\text{ V}) = \frac{1 - \text{erf}\left(\frac{1.3\bar{3}}{\sqrt{2}}\right)}{2} = 0.092 = 9.2\% \tag{4.9}$$

Manufacturers of electronic components often specify the tolerance of their parts using an absolute limit. For example, a component might have a $\pm 5\%$ tolerance meaning that the manufacturer guarantees the actual value of the part is within $\pm 5\%$ of its mean value. It is common, though not always perfectly accurate, to interpret the specification as a $\pm 3\sigma$ limit as shown in the following example.

Example 4.3. Resistors in a bin have a mean value of $4.7\,\text{k}\Omega$ and a tolerance specification of $\pm 5\%$. What is the probability of picking a resistor with a value greater than $4.85\,\text{k}\Omega$?

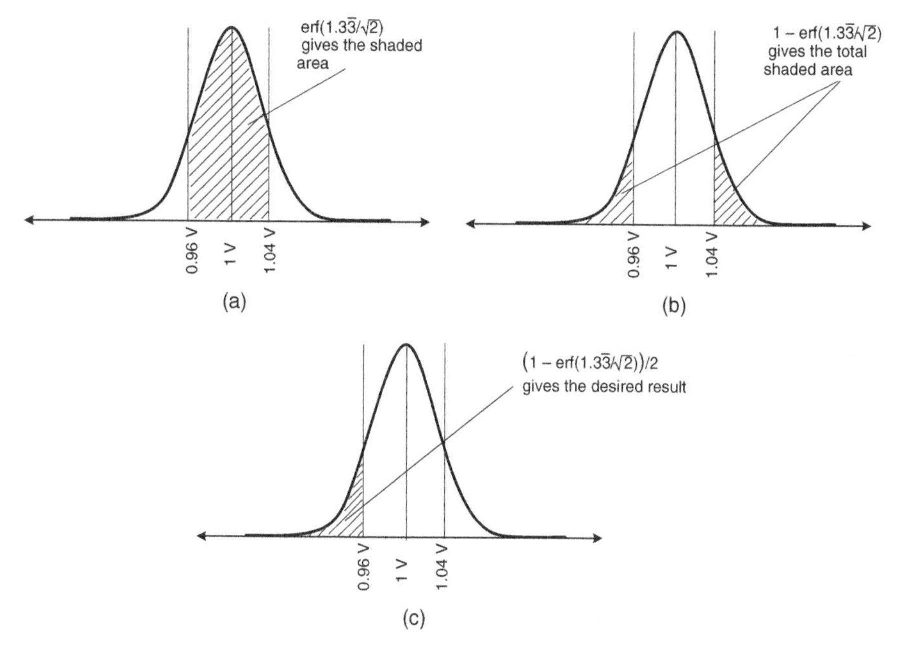

Figure 4.4. To get the probability that the measurement is less than 0.96 V start with the erf( ) function (a), then manipulate the result as shown in (b) and (c).

Solution. We interpret the $\pm 5\%$ tolerance specification as a $\pm 3\sigma$ value and compute σ:

$$\sigma = \frac{0.05}{3} \times 4700\,\Omega = 78.\bar{3}\Omega \tag{4.10}$$

The difference between the limit and the mean is normalized to σ:

$$\frac{4850\,\Omega - 4700\,\Omega}{\sigma} \times \sigma = \frac{150\,\Omega}{78.\bar{3}\,\Omega} \times \sigma = 1.91\sigma \tag{4.11}$$

The probability of picking a resistor between $\pm 1.91\sigma$ from the mean is

$$\text{erf}\left(\frac{1.91}{\sqrt{2}}\right) = 0.944 \tag{4.12}$$

The probability of picking a resistor above this range is half the probability that the resistor is outside the range in Equation 4.12:

$$P(R > 4850\,\Omega) = \frac{1 - 0.944}{2} = 0.028 = 2.8\% \tag{4.13}$$

We frequently encounter problems where we are interested in the sum of multiple variables, each from a different independent Gaussian process. This will be seen in Example 4.4 where the voltage at the output of an op-amp is affected by the input offset voltage of the op-amp and by the variations in the resistors used to set the circuit gain. Under these circumstances we use the important result that the voltage at the op-amp output is also Gaussian distributed. If N independent Gaussian-distributed variables are added, the mean and standard deviation of the result are computed as shown in Equations 4.14 and 4.15:

$$\text{mean}_{\text{Sum}} = m_1 + m_2 + \cdots + m_N \tag{4.14}$$

$$\sigma_{\text{Sum}} = \sqrt{\sigma_1^2 + \sigma_2^2 + \cdots + \sigma_N^2} \tag{4.15}$$

where m_i is the mean of the ith variable and σ_i is the standard deviation of the ith variable.

4.3 SETTING A MANUFACTURING TEST LIMIT

Consider the case where a sensor is connected to an ADC through a non-inverting amplifier[5] as shown in Figure 4.5. During manufacturing test, the path between the sensor and the amplifier is disconnected and the board test fixture injects 1 VDC at the amplifier input. The fixture then measures the amplifier output and compares it to the test limits. Provided the circuit has been properly constructed, the voltage measurement will only be affected by the resistor tolerances and the input offset voltage of the op-amp. The goal is to set the manufacturing test limits at the amplifier output.

If the limits are set too wide, then boards with incorrect or marginal components could pass the test but possibly fail in the field. If the limits are set too narrow, then correctly assembled boards with good components could fail the test, be rejected unnecessarily, and increase rework costs. Different organizations have different philosophies on how test limits should be set. Regardless of the philosophy, it is important to understand the relationship between the variations of a circuit output and the variations of the components that affect it.

Example 4.4. The tolerance of the resistors in Figure 4.5 is $\pm 2\%$ and V_{IO} for the op-amp is ± 5 mV. The tolerances can be interpreted as 3σ limits. Set the test limits at the amplifier output for a failure probability of 0.1%.

[5]Noninverting op-amp circuits are discussed in Chapter 6.

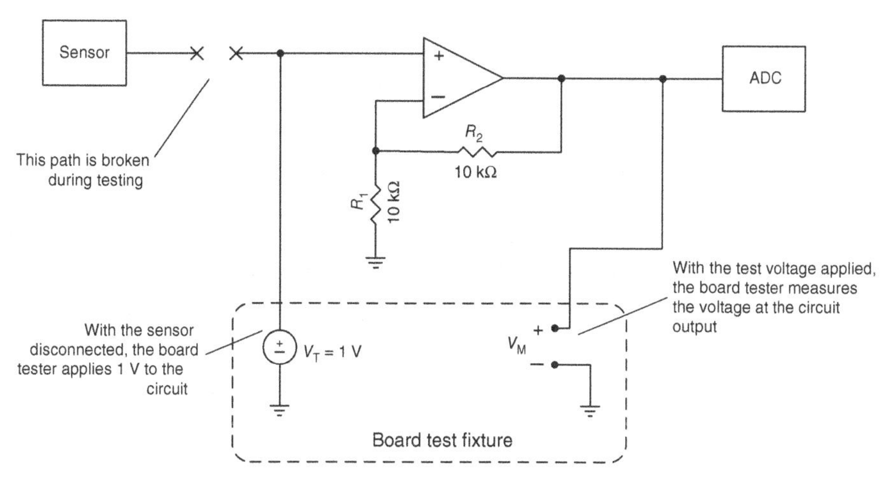

Figure 4.5. The measured voltage V_M is affected by the tolerances of R_1, R_2, and the input offset voltage of the op-amp. If the component variations are Gaussian distributed, then V_M will be Gaussian distributed also.

Solution. Our strategy is to first compute the mean value of the amplifier output. We then compute σ at the amplifier output due to the individual contributions of R_1, R_2, and V_{IO}. These contributions are combined using Equations 4.14 and 4.15 to get the standard deviation of the amplifier output. Finally we use the erf( ) function to set the test limits for the required failure probability.

In Chapter 6 it is shown that the voltage at the output of the noninverting op-amp circuit is

$$V_O = V_T\left(1 + \frac{R_2}{R_1}\right) \tag{4.16}$$

where V_T is the input voltage.

The mean of the amplifier output is the test voltage, V_T, multiplied by the gain of the circuit or

$$\text{mean} = V_T\left(1 + \frac{R_2}{R_1}\right) = 1 + 1 = 2\text{ V} \tag{4.17}$$

The change in V_O with respect to R_1 is found by differentiating Equation 4.16 with respect to R_1:

$$\frac{dV_O}{dR_1} = -V_T\frac{R_2}{R_1^2} \tag{4.18}$$

So,

$$\mathrm{d}V_{\mathrm{O}} = -\frac{R_2}{R_1^2} V_{\mathrm{T}}\, \mathrm{d}R_1 \tag{4.19}$$

The standard deviation σ at the amplifier output is computed by setting $\mathrm{d}R_1$ in Equation 4.19 equal to the standard deviation of R_1. The negative sign is dropped because the standard deviation is a statistical parameter, not a voltage:

$$\sigma_{V_{\mathrm{O}}} = \frac{R_2}{R_1^2} V_{\mathrm{T}} \sigma_{R_1} \tag{4.20}$$

Since the resistor tolerance, Tol, is a $\pm 3\sigma$ specification, σ_{R_1} is

$$\sigma_{R_1} = \frac{\mathrm{Tol}}{3} R_1 \tag{4.21}$$

Substituting Equation 4.21 into Equation 4.20 gives

$$\sigma_{V_{\mathrm{O}}} = \frac{R_2}{R_1^2} V_{\mathrm{T}} \sigma_{R_1} = \frac{R_2}{R_1^2} V_{\mathrm{T}} \frac{\mathrm{Tol}}{3} R_1 = \frac{R_2}{R_1} V_{\mathrm{T}} \frac{\mathrm{Tol}}{3} = V_{\mathrm{T}} \frac{\mathrm{Tol}}{3} \tag{4.22}$$

Substituting values into Equation 4.22 gives the standard deviation at the output due to variations in R_1:

$$\sigma_{V_{\mathrm{O}}} = V_{\mathrm{T}} \frac{\mathrm{Tol}}{3} = 1 \times \frac{0.02}{3} = 6.67\ \mathrm{mV} \tag{4.23}$$

By a similar procedure (see Problem 4.4), σ at the amplifier output due to variations of R_2 can be shown to be

$$\sigma_{V_{\mathrm{O}}} = V_{\mathrm{T}} \frac{R_2}{R_1} \frac{\mathrm{Tol}}{3} = 1 \times \frac{10,000}{10,000} \times \frac{0.02}{3} = 6.67\ \mathrm{mV} \tag{4.24}$$

In Chapter 6 it is shown that the op-amp's input offset voltage is multiplied by $1 + R_2/R_1$ and seen at the output of the op-amp. Since the $\pm 3\sigma$ tolerance for V_{IO} is 5 mV, σ at the output is[6]

$$\sigma_{V_O} = \frac{0.005}{3}\left(1 + \frac{10K}{10K}\right) = 3.33 \text{ mV} \tag{4.25}$$

The individual σ values are combined using Equation 4.15 to get σ at the amplifier output due to the three error sources[7]:

$$\sigma_{\text{Amplifier output}} = \sqrt{0.0067^2 + 0.0067^2 + 0.0033^2} = 10.00 \text{ mV} \tag{4.26}$$

The desired failure probability is 0.1% or 0.001 so the probability of success is 1 – 0.001 or 0.999. We consult an erf( ) table [1] or use the Matlab function erfinv( ) and find that erf(2.33) = 0.999. Using the scaling shown in Equation 4.5 we multiply the argument in the erf( ) function by $\sqrt{2}$ to get 3.29. Therefore our desired test limits are $\pm 3.29\ \sigma$ away from the mean or

$$\text{Test limits} = 2.0 \pm 3.29 \times 10.00 \text{ mV} \tag{4.27}$$

and the board passes the test if

$$1.967 \text{ V} < V_O < 2.033 \text{ V} \tag{4.28}$$

4.4 PROCURING A CUSTOM COMPONENT

Sometimes, instead of using off-the-shelf parts such as integrated circuits, a custom part must be specified and used. An example is the toroidal inductor shown in Figure 4.6. A toroidal inductor is a subsystem consisting of a ferrous core surrounded by a number of turns of wire. The actual inductance of each inductor depends on the core permeability and how the inductor is physically wound. Toroids are built to the designer's specifications by companies with specialized winding machines.

Say the toroid manufacturing process produces parts with a 3σ tolerance of 5% and, since the toroids are wound with an integer number of windings, a mean 2.5% above the desired value. Furthermore, we know that boards

[6] At this point we see that the contribution from V_{IO} is affected by the variations of the resistors meaning that the sources of error are not perfectly independent. Since the effect is small, we assume independence.

[7] Another useful result of this analysis is that it shows the error in the output is primarily due to the 2% resistors. Attempting to reduce the error by using an op-amp with a lower V_{IO} would increase component cost and provide only a small improvement.

Figure 4.6. Toroidal inductor. (Photo courtesy of Standex-Meder Electronics)

will fail if the parts are greater than 5% from the desired value. If the parts are installed on the boards, then manufacturing costs will rise because failures must first be traced to the toroid and then the toroid must be replaced. Sometimes toroids are measured either by the manufacturer or the board manufacturer, and parts not meeting the tolerance specification are either reworked or scrapped. In both cases manufacturing costs rise.

It is difficult to isolate and control the effects that cause manufacturing variations in subsystems such as toroidal inductors, but understanding the statistics of the process will help the designer make informed decisions about tolerance versus cost. A good way to approach the problem is to have the manufacturer build a small number of prototype samples, measure the samples, estimate statistical parameters, and finally use the estimated parameters to predict whether the parts will be acceptable when manufactured in large quantities. As shown in the next example, the Gaussian distribution provides the necessary tools.

Example 4.5. Our design requires a 6.25 μH toroidal inductor. If the inductance is not within ±5% of the mean, the design will not work properly. We must work with an inductor manufacturer to procure a part for this design.

Solution. We choose Micrometals core T94-10 [2]. The inductance of a toroid is described by Equation 4.29:

$$L = A_{\mathrm{L}} T^2 \tag{4.29}$$

TABLE 4.1. Measured inductance of 25 prototypes in μH

6.23	6.23	6.23	6.23	6.23
6.12	6.12	6.12	6.12	6.12
6.03	6.03	6.03	6.03	6.03
6.2	6.2	6.2	6.2	6.2
6.12	6.12	6.12	6.12	6.12

The measured mean is 6.08 μH. The standard deviation is 0.126 μH.

where L is the inductance in henries, A_L is the core constant, and T is the number of turns of wire wrapped around the core. The core constant for the selected core is 5.8 nH/T^2.

The number of turns required is obtained by solving Equation 4.29 for the number of turns:

$$T = \sqrt{\frac{L}{A_L}} = \sqrt{\frac{6.25 \times 10^{-6}}{5.8 \times 10^{-9}}} = 32.8\,\text{turns} \tag{4.30}$$

We procure 25 prototype inductors with the T94-10 core wrapped with 33 turns of wire. The measured inductance for the prototypes is shown in Table 4.1 and the mean and standard deviation are calculated using Excel or Matlab.

It is always worthwhile to plot a histogram [3] of measured data. The histogram shown in Figure 4.7 shows how many parts fell within each of 10 equally spaced intervals. A Gaussian pdf with the same mean and standard deviation is overlaid over the data. Since the sample size is only 25 units, it does not match the Gaussian curve exactly. But it does show that the data cluster about the mean as a Gaussian distribution should.[8] If the histogram did not exhibit this behavior it could indicate problems in the manufacturing process.[9]

The difference between the measured mean and the desired mean expressed in percent is

$$\frac{6.08\,\mu\text{H} - 6.25\,\mu\text{H}}{6.25\,\mu\text{H}} \times 100 = -2.72\% \tag{4.31}$$

[8] See [4] for methods of estimating the degree of conformance to a normal curve.

[9] If the distribution shows a hard edge corresponding to the customer's upper or lower specification then the manufacturer may be sorting parts and either discarding or reworking those that don't pass. Either the manufacturer or the customer must pay for this.

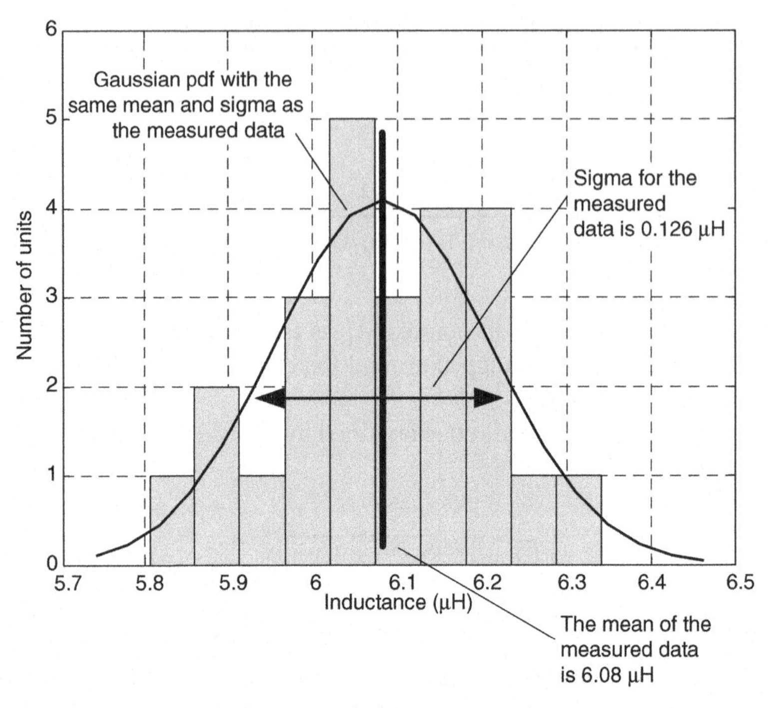

Figure 4.7. Histogram of inductance values shows how many parts fell within each of 10 equally spaced intervals. The Gaussian pdf with the measured mean and standard deviation overlaid on the histogram suggest that the data are indeed Gaussian.

The standard deviation (σ) of the measured parts expressed in percent is

$$\frac{0.126\ \mu\text{H}}{6.25\ \mu\text{H}} \times 100 = 2.02\% \tag{4.32}$$

At this point we use Figure 4.8 to evaluate the manufacturer's ability to satisfy our requirement.

The probability of these parts falling within ±5% of 6.25 μH is

$$\frac{1}{2}\left(\text{erf}\left(\frac{1.13}{\sqrt{2}}\right) + \text{erf}\left(\frac{3.82}{\sqrt{2}}\right)\right) = 87.1\% \tag{4.33}$$

The 87.1% pass rate means that about 13% of these parts will not be usable, resulting in an unacceptably high scrap rate.

Since the design formula of Equation 4.29 is only an approximation, we consider the idea of adding an integer number of turns of wire to the part. The

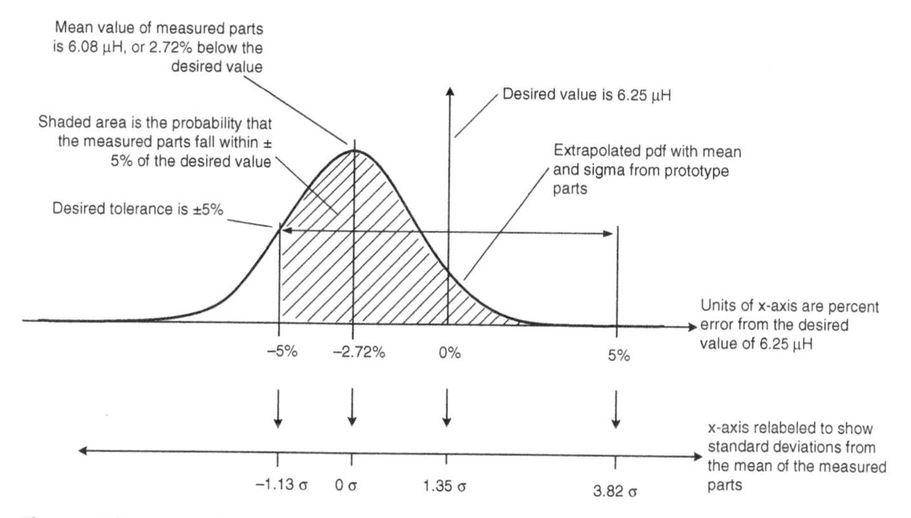

Figure 4.8. The probability of a measured part falling within ±5% of the desired value is only 87%. This is undesirable because it results in a 13% scrap rate.

percent change in inductance resulting from adding a turn of wire is

$$\begin{aligned}\%\text{Change} &= \frac{A_L\,(T+1)^2 - A_L T^2}{A_L T^2} \times 100 = \frac{(2T+1)}{T^2} \times 100 \\ &= \frac{(2.33+1)}{33^2} \times 100 = 6.15\% \qquad (4.34)\end{aligned}$$

Since Figure 4.8 shows that the error in the mean of the prototype parts is −2.72%, an adjustment of 6.15% only increases the magnitude of the error, so adjusting the number of turns is not an option.

At this point we modify our design by changing the circuitry surrounding the inductor such that the desired inductance is equal to the mean of the prototypes or 6.08 μH as shown in Figure 4.9. To determine the resulting scrap rate, we first compute the ratio of the measured σ to the mean: 0.126 μH/6.08 μH = 0.0210 = 2.1%. This makes the ±5% limit ±5/2.1 = ±2.41σ. The resulting scrap rate is

$$1 - \text{erf}\left(\frac{2.41}{\sqrt{2}}\right) = 1.6\% \qquad (4.35)$$

In addition to showing how to work with a set of measured data, this example shows the value of working cooperatively with a component vendor. In this case, the vendor could conveniently supply parts at a certain value. By

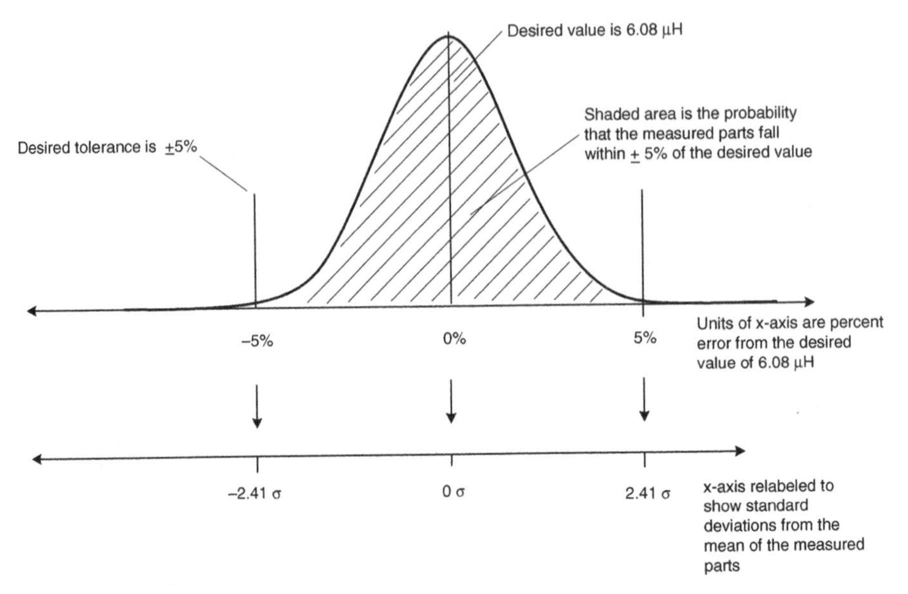

Figure 4.9. The design is changed so the desired mean matches the mean of the measured parts. This improves the scrap rate from 13% to 1.6%.

determining that value with a prototype build and then accommodating it in the design, the scrap rate for the part was reduced from 13% to 1.6%.

PROBLEMS

4.1 A radio transceiver consists of 1037 discrete components. All components are required for the radio to work properly. The probability of any component failing once the receiver is assembled is 10×10^{-6}. Assuming device failures are independent events, what is the probability that a transceiver will fail after assembly?

4.2 The relays shown in Figure 4.10 are part of a medical device. This redundant configuration is controlled so that the relays are either both open or both closed. If the path cannot be closed (meaning neither relay

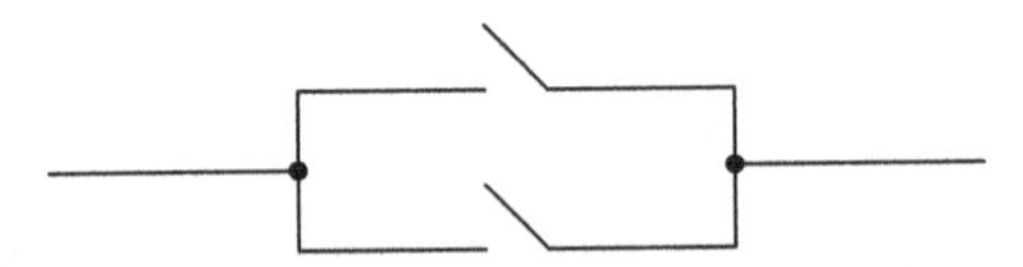

Figure 4.10. Circuit for Problem 4.2. The two relays are connected in parallel to reduce the probability of the path not closing when required.

closes) the patient's life is endangered. If the path cannot be opened (meaning neither relay opens) then the device requires replacement, but the patient is not compromised. The probability of a relay not closing or opening when required is 1×10^{-4} and relay failures are considered independent. What is the probability that the path will not close when required? What is the probability that the path will not open when required?

4.3 The mean of a Gaussian-distributed process is 48.22 and the standard deviation is 4.0. Use the erf( ) function to determine the probability that a value would fall in the range $49.1 \leq x \leq 50.1$.

4.4 In Example 4.4 the standard deviation of the voltage at the amplifier output due to R_1 was $\sigma_{V_O} = V_O(\mathrm{Tol}/3)$. Show that the standard deviation at the amplifier output due to R_2 is $\sigma_{V_O} = V_T(R_2/R_1) \cdot (\mathrm{Tol}/3)$.

4.5 A test point in a circuit is affected by 50 Gaussian-distributed sources of error. The standard deviation from each error source at the test point is 10 mV. The mean voltage at the circuit output is 0 V. What should the test limits be so that boards within $\pm 3\sigma$ will pass? Another engineer suggests that since the "worst case" from each source is 3σ or 30 mV, the worst-case test limits should be 50 times that or ± 1.5 V. What do you tell him?

REFERENCES

1. Wikipedia. *Error Function (and erf Table).* Available: http://en.wikipedia.org/wiki/Error_function (Accessed 11-30-2013.)
2. Micrometals Corporation. *RF Applications Catalog.* Available: http://www.micrometals.com/catalog_index.html (Accessed 11-30-2013.)
3. Wikipedia. *Histogram.* Available: http://en.wikipedia.org/wiki/Histogram (Accessed 11-30-2013.)
4. R. H. Montgomery, *Engineering Statistics*, John Wiley and Sons, 2011.

5

HOW TO DESIGN A FEEDBACK CONTROL SYSTEM

Feedback control systems allow a person or a machine to specify something a machine must do. For example, a surgeon can use a robotically controlled instrument to perform precision surgery, or a radio receiver can adjust its gain so the listener hears the desired volume despite the received signal level varying by decades of magnitude. Or on the manufacturing floor, a cutting wheel can be controlled to make precise cuts even though the characteristics of the blade change as it wears. Another use for control systems is self-calibration of electronic systems. In these applications an accurate sensor is used to compensate for manufacturing variations, temperature variations, and aging of system components. This not only reduces manufacturing costs, but also insures a long product lifetime.

The above examples show that control systems, or servos,[1] save human effort, reduce manufacturing costs, and increase product quality, and they all include electronics, so control systems should unquestionably be a part of

[1]The terms "control system" and "servo" are used interchangeably in this chapter.

Ten Essential Skills for Electrical Engineers, First Edition. Barry L. Dorr.

your skill set. Anyone who has designed, debugged, and then watched a servo in operation will tell you that it was fascinating and enjoyable. Developing a servo often involves theory, circuitry, firmware, heat transfer, and mechanical parts. The best servo designers put on their safety glasses and then bend things, break things, and sometimes get firsthand experience with the fire extinguisher!

Analysis and design of sophisticated servo systems can quickly become complicated and difficult, but the majority of servo design requirements can be satisfied using the simple first- and second-order servos discussed in this chapter. We begin by showing that most servo systems you encounter can be put in the classical block diagram form shown in control systems textbooks and analyzed using a few basic equations.[2] We then discuss the parameters typically used to quantify or specify the performance of servo systems. Next, we discuss why first- and second-order servos are so common, and then show how to design them. For those that would like to gain some hands-on experience, Section 5.5 shows a schematic diagram of a simple second-order servo which can be built from a few inexpensive parts. Finally we show how to design discrete servo controllers using hardware or firmware.

5.1 INTUITIVE DESCRIPTION OF A CONTROL SYSTEM

Consider the automotive speed control system in Figure 5.1. This system accepts a desired speed from the driver and controls the accelerator so the car will travel at the desired speed. Intuition tells us that pressing the accelerator harder generally makes the vehicle go faster, but we know that that there is not a one-to-one correspondence between accelerator position and speed. In fact, while driving we usually have no idea how far the accelerator is from the floor. The uncertainty is partly due to the engine and drivetrain and partly due to the terrain profile, that is, whether the car is traveling uphill or downhill. The path between the accelerator position and the vehicle speed is modeled in Figure 5.1 by the *actuator*, which represents the engine/drivetrain, and a summing junction that represents how the car's speed is affected by terrain (the *disturbance*). The vehicle speed is measured by a *sensor* in the system and fed to the controller that compares the measured speed to the desired speed or *setpoint* by subtracting the former from the latter. The result of the subtraction is the *error* signal which is filtered and fed to the accelerator pedal.

[2]This is particularly important for job interviews.

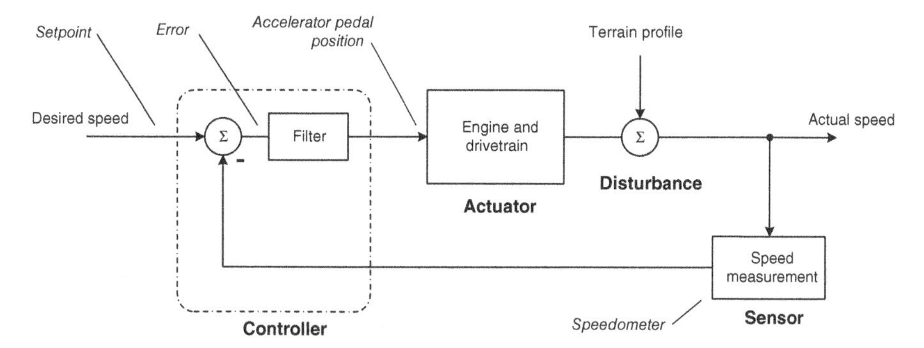

Figure 5.1. The basic elements of a control system are shown in an automotive speed control system.

In the example above, the servo system frees a human operator from having to perform the critical but simple task of speed control. Another common motivation for using a servo system is to use an accurate sensor to control the output of an actuator which is corrupted by disturbances or other inaccuracies. In many cases the actuator is the cornerstone of the servo design because it is the most expensive component.

5.2 REVIEW OF CONTROL SYSTEM OPERATION

The key to understanding the control system of Figure 5.2 is to note that it always attempts to drive the error signal to zero, which is the same as driving the system output to match the setpoint. If a disturbance makes the output signal less than the setpoint, then the error signal will be positive and the

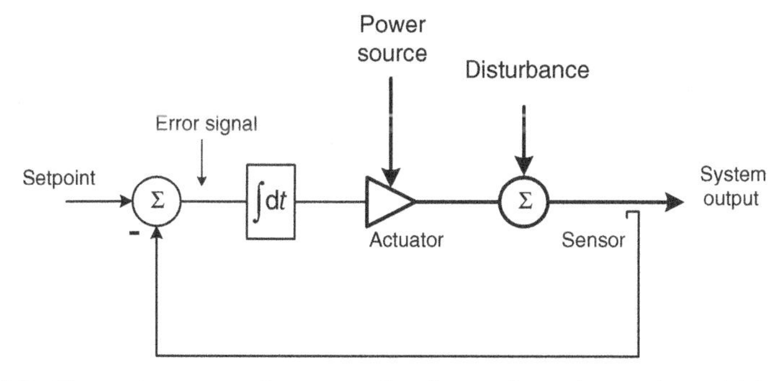

Figure 5.2. Elementary control system. The heavy lines show where high power is flowing.

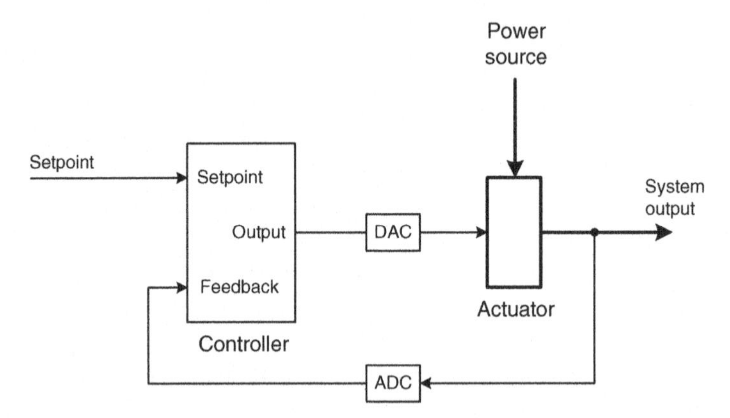

Figure 5.3. Showing controller/actuator viewpoint. The controller is frequently a small, stand-alone unit consisting of analog or digital circuitry.

integrator[3] output will increase, thereby causing the actuator and hence the system output to increase. If the output is greater than the setpoint, then the error signal will be negative and the integrator output will decrease, causing the actuator and the system output to decrease.

The bold lines in Figure 5.2 emphasize the notion that the actuator is frequently the device that manipulates the power in the system. It moves the load, controls the current, heats the room, and so on. The disturbance illustrates the idea that the transfer function between the actuator input and system output can change with varying operating conditions. For example, if the actuator lifts a load, the weight of the load will affect its input–output relationship. When the system is observed during operation, a disturbance will initially cause an error that will cause the integrator output to change, thereby adjusting the actuator so that its output again matches the setpoint, which causes the error to return to zero. Note that when the disturbance is corrected, the integrator has changed but the system output has not.

The feedback control system of Figure 5.2 is often represented as shown in Figure 5.3. This representation emphasizes the controller as a stand-alone unit and is often seen in commercial literature.

For analysis and design of control systems, a system schematic such as the one shown in Figure 5.4a is converted to the simple block diagram shown in Figure 5.4b.

[3]The filter of Figure 5.1 is replaced by an integrator in Figure 5.2. Using an integrator guarantees that the *steady-state error* is zero when the setpoint is held constant. All of the systems in this chapter use an integrator as a filter.

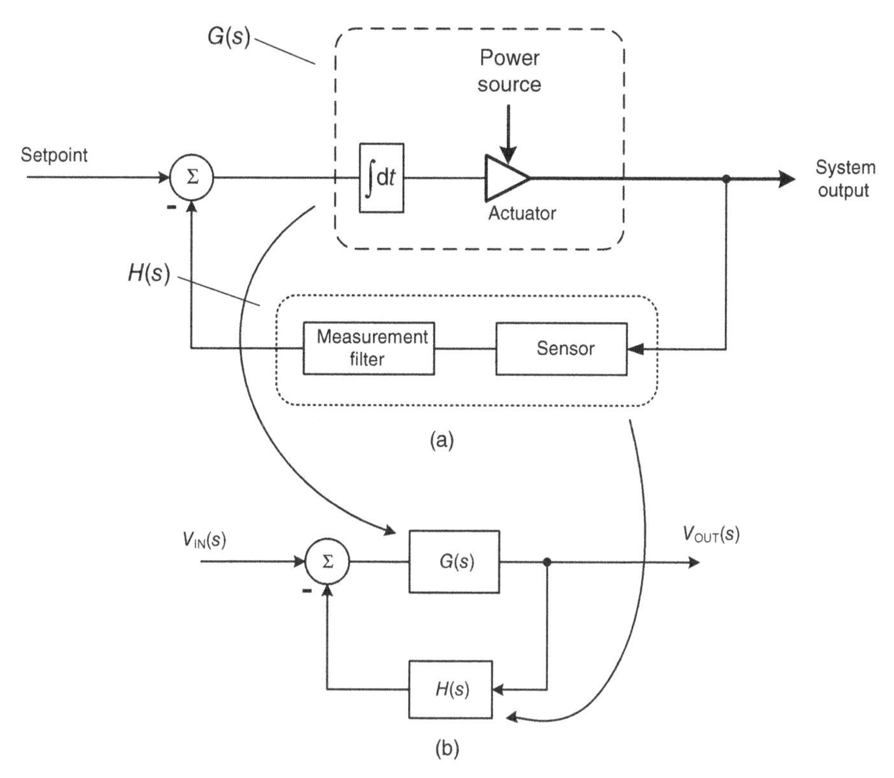

Figure 5.4. A schematic of a control system (a) is represented by the classical system diagram (b). The classical system diagram is applicable to any control system.

The most common transfer function for a control system is the ratio of the system output to the system input in the frequency domain. It is derived by inspection of Figure 5.4b which yields the following equation:

$$[V_{\mathrm{IN}}(s) - V_{\mathrm{OUT}}(s) \cdot H(s)]G(s) = V_{\mathrm{OUT}}(s) \tag{5.1}$$

where $G(s)$ is the forward transfer function and $H(s)$ is the reverse transfer function.

Equation 5.1 can be rearranged to give the classical transfer function for the closed-loop response[4]:

$$\frac{V_{\mathrm{OUT}}}{V_{\mathrm{IN}}}(s) = T(s) = \frac{G(s)}{1 + G(s) \cdot H(s)} \tag{5.2}$$

[4]Any control system, simple or complex, can be reduced to this basic form.

5.3 PERFORMANCE OF CONTROL SYSTEMS

When evaluating or designing a control system, the following points should always be considered:

Stability—Does the feedback loop remain stable under all expected conditions including component drift, temperature drift, and manufacturing variability of the components used[5]?

Step response—How long does it take for the system to respond to an abrupt change in the setpoint or a disturbance? Does it overshoot its target value?

Steady-state error—When the system has settled, does the error signal go to zero[6]?

Dynamic range—Are the system components capable of responding to the expected system excursions? If the system is electrical does it have sufficient supply voltage? If it is mechanical will something bend or break?

5.4 FIRST-ORDER CONTROL SYSTEM DESIGN

Many control system design requirements do not require the system to respond quickly to a disturbance or change in the setpoint. When this is the case, the response time of the system will be much longer than the response time of the actuator and sensor. Under these conditions the first-order servo described in this section is an excellent solution because of the following:

1. It has no steady-state error to a step input.
2. It is unconditionally stable.
3. It will not overshoot in response to a step input.
4. Its closed-loop transfer function is the same as a lowpass RC filter thereby simplifying analysis.
5. It can easily be implemented in hardware or as code in a microcontroller, as shown in Section 5.7.

The first-order servo is shown in Figure 5.5. Since the response time of the system is significantly longer than the response time of the actuator, we ignore

[5]Stability is a major concern for control systems. The systems presented in this chapter are all unconditionally stable so it is not discussed. See Chapter 6 for more on stability.

[6]If the system has a single integrator it will have no steady-state error in response to a fixed setpoint once the system settles.

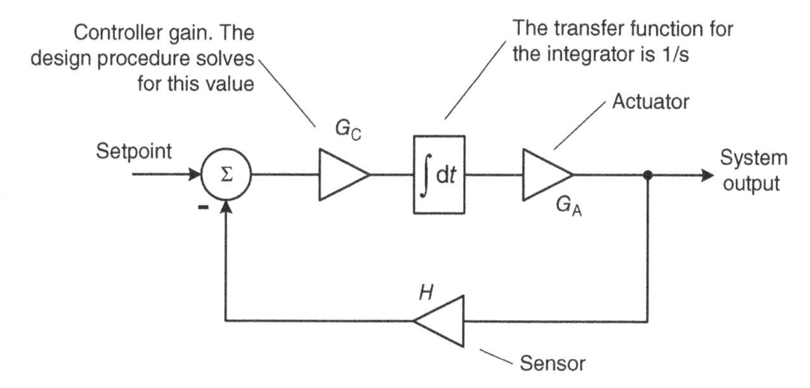

Figure 5.5. First-order servo. This topology is used when the required system response time is much greater than the response time of the actuator and sensor.

the frequency response characteristics of the actuator and represent it as gain value G_A rather than $G(s)$ as it was shown in Equation 5.2. Similarly, we also ignore the frequency response characteristics of the sensor and represent it as a constant gain of H. The constant G_C is the gain that sets the system response.

The transfer function for the system of Figure 5.5 is determined using Equation 5.2:

$$\frac{V_{OUT}}{V_{IN}}(s) = \frac{G(s)}{1 + G(s)H(s)} = \frac{G_C \cdot \frac{1}{s} \cdot G_A}{1 + G_C \cdot \frac{1}{s} \cdot G_A H} = G_C G_A \frac{1}{s + G_C G_A H} \tag{5.3}$$

From Figure 5.5 we note that the product $G_C G_A H$ is the total fixed gain in the loop. We call this G_L and rewrite Equation 5.3 as

$$\frac{V_{OUT}}{V_{IN}}(s) = \frac{1}{H} G_L \cdot \frac{1}{s + G_L} \tag{5.4}$$

Defining

$$\tau = \frac{1}{G_C G_A H} = \frac{1}{G_L} \tag{5.5}$$

Equation 5.4 is rewritten to give

$$\frac{V_{OUT}}{V_{IN}}(s) = \frac{1}{H} \cdot \frac{1}{\tau} \cdot \frac{1}{s + \frac{1}{\tau}} \tag{5.6}$$

which is identical to the transfer function for a lowpass RC filter multiplied by a scaling factor of $1/H$.[7]

Equation 5.6 makes intuitive sense by considering the system of Figure 5.5 when the input is constant and the output has settled to the steady-state value. Under these conditions the output of the sensor is equal to the setpoint. Working backwards through the sensor, the system output is the setpoint multiplied by $1/H$. The steady-state response of Equation 5.6 is computed using asymptotic analysis by substituting $s = 0$ with the result that the system output is $1/H$ times the setpoint.

The analysis is simplified if $H = 1$ which conveniently sets the output equal to the setpoint at steady state. Equation 5.6 becomes

$$\frac{V_{\mathrm{OUT}}}{V_{\mathrm{IN}}}(s) = \frac{1}{\tau} \cdot \frac{1}{s + \frac{1}{\tau}} \tag{5.7}$$

The step response for the system is

$$V_{\mathrm{O}}(t) = V_{\mathrm{F}} - (V_{\mathrm{F}} - V_{\mathrm{I}})\mathrm{e}^{-t/\tau} \tag{5.8}$$

where V_{F} is the final output voltage at $t = \infty$ and V_{I} is the initial output voltage. Substituting $V_{\mathrm{F}} = V_{\mathrm{IN}}$[8] and $V_{\mathrm{I}} = 0$, Equation 5.8 becomes

$$V_{\mathrm{O}}(t) = V_{\mathrm{IN}}(1 - \mathrm{e}^{-t/\tau}) \tag{5.9}$$

To summarize, if the system design requirements can be met with a response time significantly longer than the response time of the actuator and sensor, then the loop gain is computed from Equation 5.5:

$$G_{\mathrm{L}} = \frac{1}{\tau} \tag{5.10}$$

and the system will have the exponential response of Equation 5.9. This illustrates the valuable result that increasing the gain of a first-order system speeds up the response time.[9]

[7] See Chapter 7 for design and analysis of the lowpass RC filter.

[8] The output voltage at $t = \infty$ is equal to the input voltage. Visualize the lowpass RC filter.

[9] Provided the reduced response time is still much longer than the response time of the actuator and sensor!

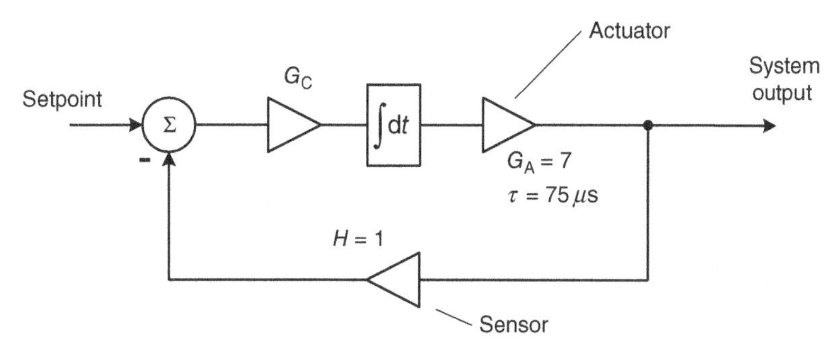

Figure 5.6. System for Example 5.1.

Example 5.1. The DC gain of the actuator in Figure 5.6 is 7 and its time constant 75 μs. The DC gain of the sensor is unity. Determine the value for G_C so that the control system will respond to a step input by reaching 98% of the final output within 5 ms. Plot the step response.

Solution. The required system response time is significantly longer than the actuator's time constant so the actuator can be considered a multiplication by 7 and Equation 5.9 describes the step response. First the response time and required percentage of the final value at the response time are substituted into Equation 5.9:

$$0.98 = 1\left(1 - e^{-0.005/\tau}\right) \tag{5.11}$$

Solving for time constant τ gives

$$\tau = \frac{-0.005}{\ln(1 - 0.98)} = 1.28\ \text{ms} \tag{5.12}$$

The loop gain is computed using Equation 5.10:

$$G_L = \frac{1}{\tau} = \frac{1}{0.00128} = 782 \tag{5.13}$$

The controller gain is computed by setting the product of all loop gains to G_L:

$$G_C = \frac{G_L}{G_A H} = \frac{781}{7 \times 1} = 112 \tag{5.14}$$

The step response is plotted using Equation 5.9 and shown in Figure 5.7.

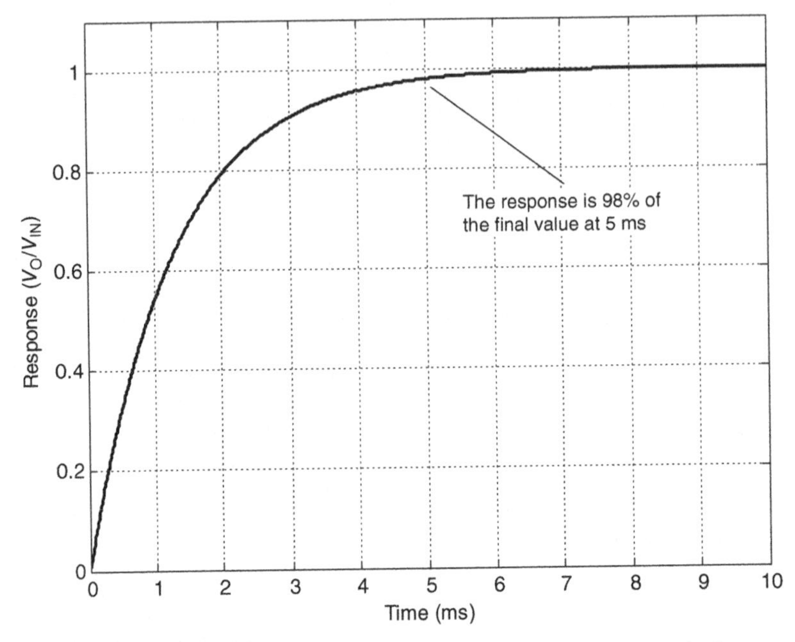

Figure 5.7. Step response for system of Example 5.1. The step response is the same as a lowpass RC filter.

5.5 SECOND-ORDER CONTROL SYSTEM DESIGN

The control system described in Section 5.4 was simple because the actuator was represented by a gain block rather than a frequency-dependent transfer function. This was possible because the required response time was significantly longer than the response time of the actuator and sensor. This section covers the case where maximum performance from an actuator is required, that is, when system performance is limited by the actuator's response time.

The frequency response of many electrical, mechanical, or thermal actuators is maximum at DC and falls off with increasing frequency. It is common to approximate these actuators using the transfer function for a lowpass RC filter. This results in the system shown in Figure 5.8. Many practical control systems use this topology because of the following:

1. The integrator eliminates any steady-state error for a step input.
2. The sluggishness of the actuator is represented as a single-pole filter, that is, a lowpass RC filter.
3. It is unconditionally stable.
4. Precompiled design material is available to assist with the design process.

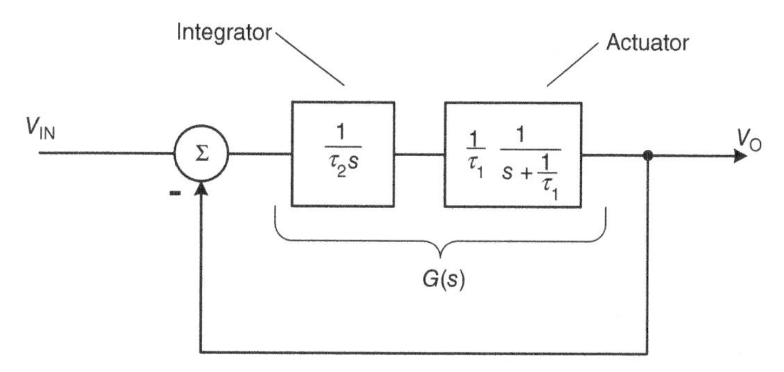

Figure 5.8. System with an integrator and frequency-dependent actuator. The actuator is modeled as a lowpass RC filter.

In Figure 5.8 $H(s) = 1$, so the transfer function of Equation 5.2 simplifies to

$$T(s) = \frac{G(s)}{1 + G(s)} \tag{5.15}$$

The equation for $G(s)$ is that of the integrator followed by the actuator:

$$G(s) = \frac{1}{\tau_2 s} \cdot \frac{1}{\tau_1} \cdot \frac{1}{s + \frac{1}{\tau_1}} \tag{5.16}$$

Substituting Equation 5.16 into Equation 5.15 gives the second-order transfer function

$$T(s) = \frac{1}{\tau_1 \tau_2} \cdot \frac{1}{s^2 + \frac{1}{\tau_1} s + \frac{1}{\tau_1 \tau_2}} \tag{5.17}$$

It is worthwhile to relate Equation 5.17 to the general second-order transfer function as shown in Equation 5.18 because there are many useful formulas that relate performance parameters such as response time, overshoot, and bandwidth of second-order systems to the constants ω_n and ξ [1]:

$$T(s) = \frac{1}{\tau_1 \tau_2} \cdot \frac{1}{s^2 + \frac{1}{\tau_1} s + \frac{1}{\tau_1 \tau_2}} = \frac{\omega_n^2}{s^2 + 2\xi\omega_n s + \omega_n^2} \tag{5.18}$$

Equating coefficients gives

$$\omega_n = \frac{1}{\sqrt{\tau_1 \tau_2}} \tag{5.19}$$

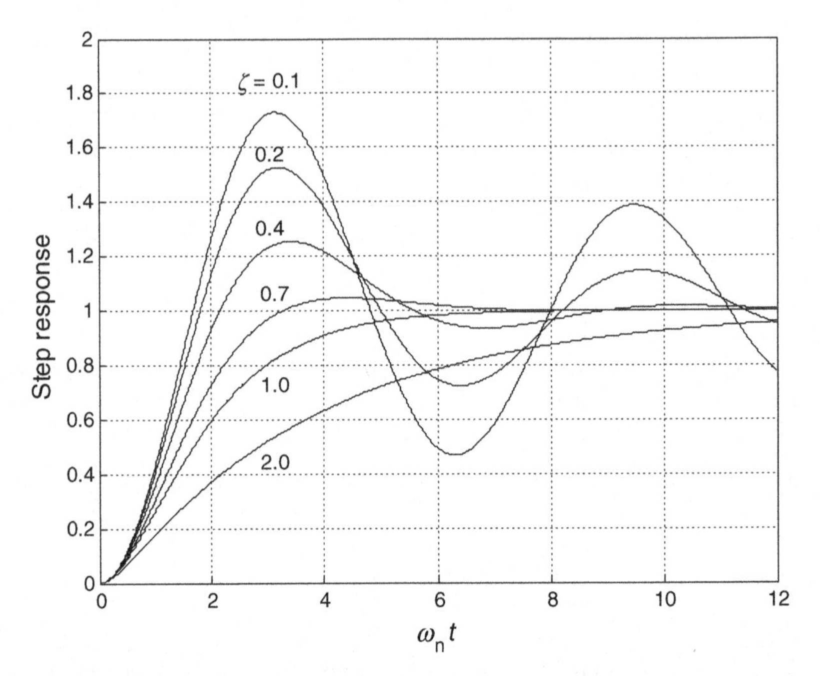

Figure 5.9. Normalized step response for the second-order control system of Figure 5.8.

and

$$\xi = \frac{1}{2}\sqrt{\frac{\tau_2}{\tau_1}} \tag{5.20}$$

ω_n is called the *natural frequency* and ξ is the *damping ratio.*

The step response of the general second-order system can be computed by multiplying the expression on the right side of Equation 5.18 by the Laplace transform of a unit step, or $1/s$, and then computing the inverse Laplace transform. This is plotted in Figure 5.9.

Figure 5.9 shows that if the damping ratio, ξ, is less than 1.0 then the system will have overshoot. Furthermore, when the system has overshoot, the following formulas apply.

The settling time, defined as four time constants, for a step input is

$$T_S = \frac{4}{\xi\omega_n}\text{s} \tag{5.21}$$

The time to reach the first peak is

$$T_P = \frac{\pi}{\omega_n} \cdot \frac{1}{\sqrt{1-\xi^2}}\text{s} \tag{5.22}$$

And the percent overshoot is

$$\text{P.O.} = 100\text{e}^{-\xi(\pi/\sqrt{1-\xi^2})}\% \tag{5.23}$$

Example 5.2. Say the servo system in Figure 5.8 has $\tau_1 = 87\ \mu\text{s}$ and $\tau_2 = 290\ \mu\text{s}$. Determine the damping ratio and settling time for the system.

Solution. First compute ω_n using Equation 5.19:

$$\omega_n = \frac{1}{\sqrt{\tau_1\tau_2}} = \frac{1}{\sqrt{87\times10^{-6}\times290\times10^{-6}}} = 6.30\times10^3\text{rad/s} \tag{5.24}$$

Then compute ξ using Equation 5.20:

$$\xi = \frac{1}{2}\sqrt{\frac{\tau_2}{\tau_1}} = \frac{1}{2}\sqrt{\frac{290\times10^{-6}}{87\times10^{-6}}} = 0.913 \tag{5.25}$$

The settling time is computed using Equation 5.21:

$$T_S = \frac{4}{\xi\omega_n} = \frac{4}{0.913\times6.30\times10^3} = 696\ \mu\text{s} \tag{5.26}$$

To analyze the second-order control system of Example 5.2 we solved for ω_n using Equation 5.19 and solved for ξ using Equation 5.20. Precompiled design information from [1] was then used to determine the behavior of the system. Design of a second-order control system can be done using the step response plot Figure 5.9 as a guide. For design purposes, Equations 5.19 and 5.20 have been manipulated into the forms below which allow computation of τ_1 and τ_2 based on design inputs ω_n and ξ:

$$\tau_1 = \frac{1}{2\xi\omega_n} \tag{5.27}$$

$$\tau_2 = \frac{1}{\omega_n^2\tau_1} \tag{5.28}$$

Example 5.3. Given an actuator with $\tau_1 = 12$ ms, design a control system with a damping ratio of 0.7. Determine the settling time, time to first peak, and percent overshoot when a step input is applied, and plot the step response.

Solution. Since τ_1 and ξ are specified, the design process consists of computing τ_2. First, Equation 5.27 is solved for ω_n:

$$\omega_n = \frac{1}{2\xi\tau_1} = \frac{1}{2 \times 0.7 \times 0.012} = 59.5\ \text{rad/s} \tag{5.29}$$

Next ω_n and τ_1 are substituted into Equation 5.28 and τ_2 is computed:

$$\tau_2 = \frac{1}{\omega_n^2\tau_1} = \frac{1}{59.52^2 \times 0.012} = 23.5\ \text{ms} \tag{5.30}$$

The settling time of the system is computed from Equation 5.21:

$$T_S = \frac{4}{\xi\omega_n} = \frac{4}{0.7 \times 59.5} = 96.0\ \text{ms} \tag{5.31}$$

The time to reach the first peak is computed from Equation 5.22:

$$T_P = \frac{\pi}{\omega_n}\frac{1}{\sqrt{1-\xi^2}} = \frac{\pi}{59.5}\frac{1}{\sqrt{1-0.7^2}} = 73.9\ \text{ms} \tag{5.32}$$

And the percent overshoot is computed from Equation 5.23:

$$\text{P.O.} = 100\text{e}^{-\xi(\pi/\sqrt{1-\xi^2})} = 100\text{e}^{-0.7(\pi/\sqrt{1-0.7^2})} = 4.59\% \tag{5.33}$$

The step response for the system is observed by dividing the horizontal axis of Figure 5.9 repeated as Figure 5.10 by ω_n and using the curve for $\xi = 0.7$.

In Example 5.3 we saw that a damping ratio of 0.7 provided a trade-off between a large overshoot resulting from a low damping ratio and the sluggish response from a high damping ratio. If the actuator gain tends to vary,[10] then the dynamic loop response will vary as shown in the following example.

[10]Varying actuator gain is a frequent motivation for using a control system in the first place!

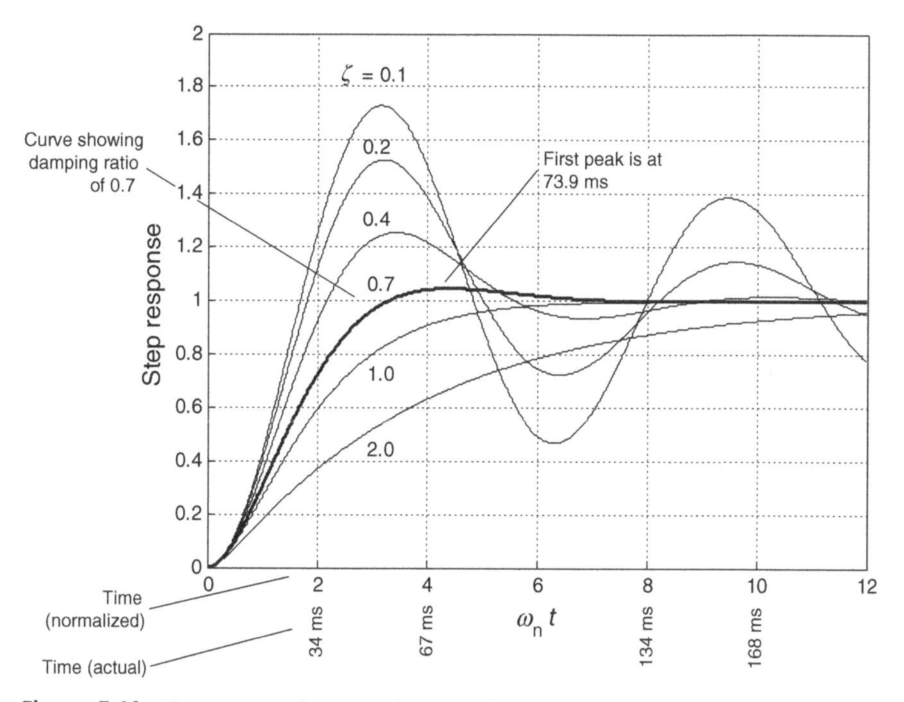

Figure 5.10. The response for Example 5.3 is shown by denormalizing the generalized curve of Figure 5.9.

Example 5.4. Determine the percent overshoot of the system of Example 5.3 if the actuator gain is doubled.

Solution. The forward path gain of the original system is

$$G(s) = \frac{1}{\tau_2 s} \cdot \frac{1}{\tau_1} \cdot \frac{1}{s + \frac{1}{\tau_1}} \tag{5.34}$$

Referring to Figure 5.8 or Equation 5.34 we can see that dividing τ_2 by 2 doubles the loop gain. Substituting 11.75 ms for τ_2 in Equation 5.19 gives

$$\omega_n = \frac{1}{\sqrt{\tau_1 \tau_2}} = \frac{1}{\sqrt{0.012 \times 0.01175}} = 84.2 \text{ rad/s} \tag{5.35}$$

Next solve Equation 5.27 for ξ:

$$\xi = \frac{1}{2\tau_1 \omega_n} = \frac{1}{2 \times 0.012 \times 84.2} = 0.495 \tag{5.36}$$

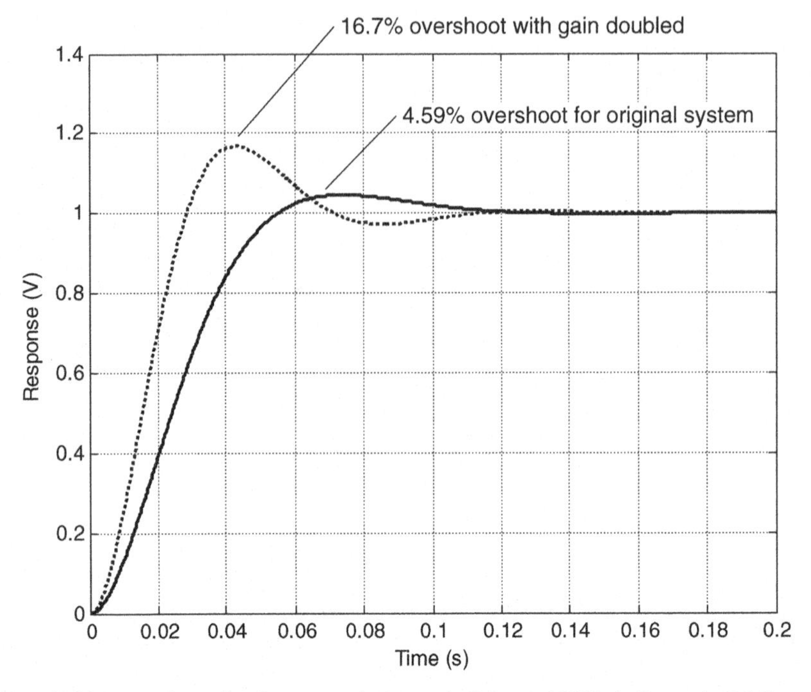

Figure 5.11. Overshoot for the system in Example 5.3 was 4.59%. In Example 5.4 the gain was doubled raising the overshoot to 16.7%.

Finally use Equation 5.23 to compute the percent overshoot:

$$\text{P.O.} = 100e^{-\xi(\pi/\sqrt{1-\xi^2})} = 100e^{-0.495(\pi/\sqrt{1-0.495^2})} = 16.7\% \tag{5.37}$$

Figure 5.11 shows the system step response for the original system of Example 5.3 and for the system with the gain doubled.

5.6 CIRCUIT REALIZATION OF A SECOND-ORDER CONTROL SYSTEM

Figure 5.12 shows a circuit that implements the second-order system from Example 5.3 and can be easily built using inexpensive parts.[11] This circuit will give you valuable experience working with a second-order servo. Time constants for the circuit are $\tau_1 = 12$ ms and $\tau_2 = 23.5$ ms.

[11]Companies such as Digi-Key sell parts in small quantities [2].

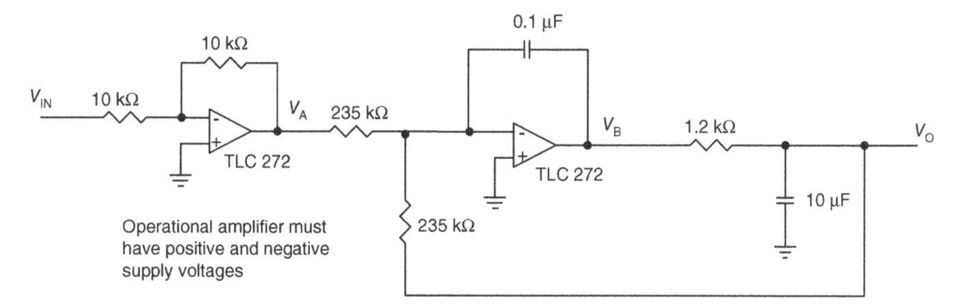

Figure 5.12. Circuit realization of second-order control system. Building or simulating this circuit will provide valuable experience with second-order servos.

The op-amp on the right provides the functions of both summing node and integrator.[12] Since the integrating op-amp inverts, connecting the system output to the 235 kΩ resistor provides the required subtraction at the summing node. The op-amp on the left is an inverter that realizes the function

$$V_A = -V_{IN} \tag{5.38}$$

The output of the integrator is

$$\begin{aligned} V_B &= -\left(V_A + V_O\right)\frac{1}{235k\Omega \times 0.1\ \mu\text{F} \times s} \\ &= -\left(V_A + V_O\right)\frac{1}{23.5\ \text{ms} \times s} = -\left(V_A + V_O\right)\frac{1}{\tau_2 s} \end{aligned} \tag{5.39}$$

Substituting Equation 5.38 into Equation 5.39 shows that the integrator integrates the error:

$$V_B = \left(V_{IN} - V_O\right)\frac{1}{\tau_2 s} \tag{5.40}$$

The resistor and capacitor at the right side of the circuit model the actuator time constant of 12 ms. A ceramic dielectric is a good choice for the 10 μF capacitor because it must not be polarized.

5.7 FIRST-ORDER DISCRETE CONTROL SYSTEM

The control systems discussed thus far have been analog systems as shown in Figure 5.13a. It is more common to find systems implemented with analog

[12] See Chapter 6 for a discussion of op-amp integrators.

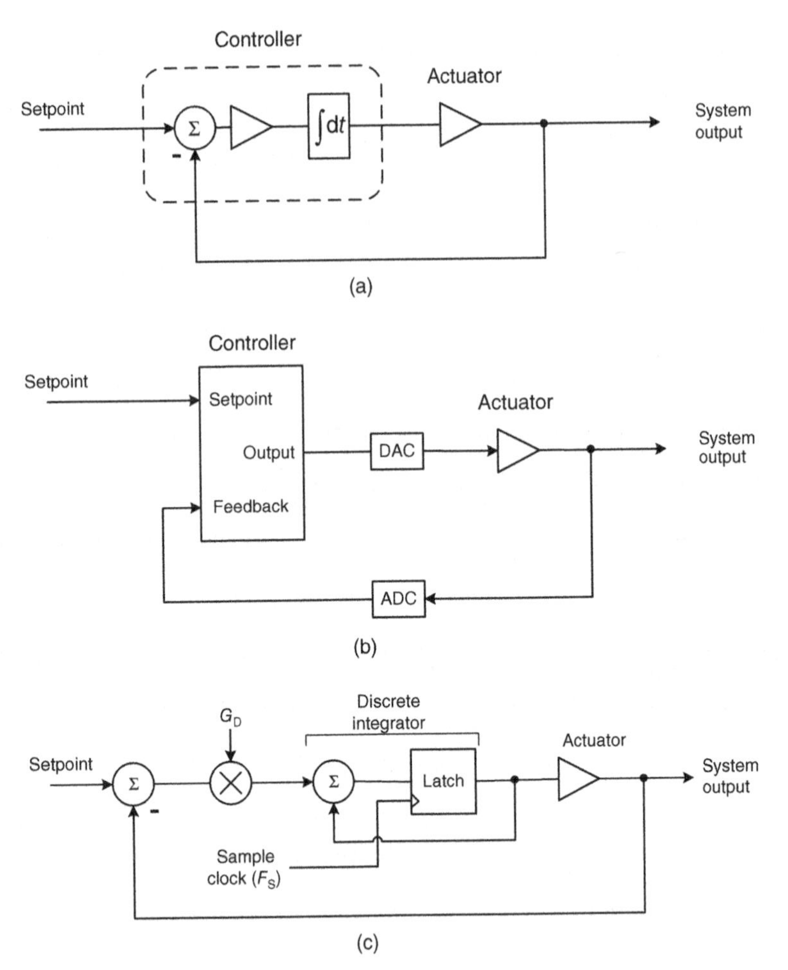

Figure 5.13. The analog system in (a) can be implemented in firmware or digital hardware as shown in (b). The analysis/design model is shown in (c).

actuators controlled by firmware or digital hardware as shown in Figure 5.13b. This section shows how to design a discrete controller based on the first-order system described in Section 5.4. The method is applicable when the discrete system sampling rate can be set much higher than the bandwidth of the closed-loop system. We use a factor of about 50,[13] but if the factor is less, the discrete system will often work satisfactorily even though the response of the discrete and analog systems will not exactly match.

[13]The Nyquist sampling criteria allows a sampling rate as low as twice the system bandwidth. The factor of 50 is used to make the discrete system match the analog prototype. Oversampling is frequently practical in microcontrollers and digital hardware. These concepts are also discussed in Chapter 8.

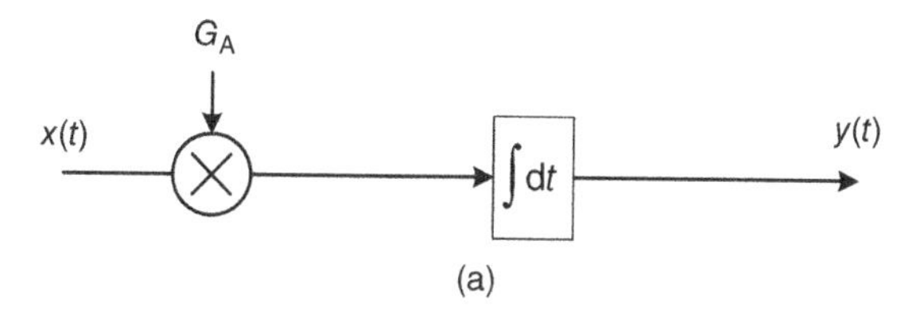

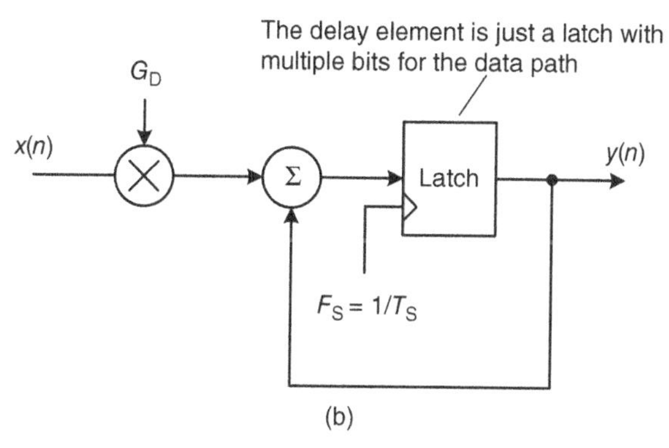

Figure 5.14. Analog (a) and discrete (b) integrators. As the sample rate increases the two behave more alike.

The controllers discussed in this section include a gain block followed by a discrete integrator as shown in Figure 5.13c. To design the discrete system, we first examine relationships between the analog and discrete integrators. The analog integrator is shown in Figure 5.14a; the discrete integrator is shown in Figure 5.14b.

To get an intuitive feel for the operation of the discrete integrator as well as its limitations, consider the case where the inputs to both integrators in Figure 5.14 are held constant during one sampling period. For an input of unity, the output of the analog integrator will increase by

$$\Delta_A = G_A \int_0^{T_S} 1\,dt = G_A T_S \tag{5.41}$$

where T_S is the sample period which is the reciprocal of the sample rate F_S.

The output of the discrete integrator will increase by

$$\Delta_D = G_D \tag{5.42}$$

So if we set

$$G_D = G_A T_S \tag{5.43}$$

then the analog and discrete integrators would produce identical results.

Now consider the case where the input varies during the sample period. The analog integrator would accurately integrate the signal, but the discrete integrator would simply add the current sample to the last sample thereby incurring error. A higher sampling rate will result in less input signal variation during a sample period. This is why a high sample rate insures a better match between the two integrators.[14]

We can now design the discrete controller. Since the controller is first order, it is fully described by its time constant or equivalently by its 3 dB bandwidth. The procedure is shown in the steps below:

1. Select a sample rate approximately 50 times the loop bandwidth or higher.
2. Determine G_A, the loop gain required for an analog implementation using Equation 5.10 and repeated as Equation 5.45.
3. Determine the value of gain for the discrete system, G_D, using Equation 5.43. This is the total gain around the loop.
4. Adjust G_D to accommodate the gain of the DAC, ADC, and actuator.

The time constant of the system of Figure 5.13a is related to its bandwidth by

$$\mathrm{BW} = \frac{1}{2\pi\tau} \tag{5.44}$$

Furthermore, from Equation 5.10 the total loop gain for an analog implementation is

$$G_A = \frac{1}{\tau} = 2\pi\,\mathrm{BW} \tag{5.45}$$

where G_A is the total gain around the loop. Note that this would include the gain of the actuator and the sensor.

The gain of the discrete system is computed using Equation 5.43:

$$G_D = G_A T_S = \frac{1}{\tau} T_S = \frac{1}{\tau F_S} \tag{5.46}$$

Example 5.5. Design a discrete controller with time constant 2 ms based on the topology of Figure 5.13c. The product of the ADC, DAC, and actuator

[14]This technique is referred to as the *method of backward differences* and can be used to convert other analog structures also. See Chapter 8.

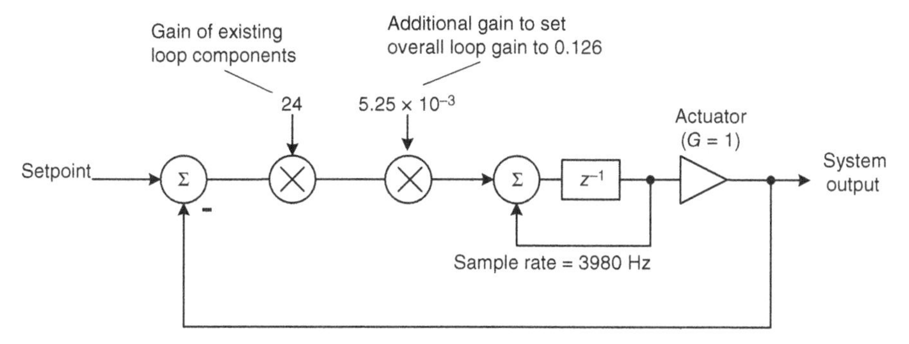

Figure 5.15. Discrete implementation of the system in Example 5.5.

gains is 24. Determine the required gain for the discrete controller and compare the closed-loop step response to an analog system with the same time constant.

Solution. The system bandwidth is determined using Equation 5.44:

$$\mathrm{BW} = \frac{1}{2\pi\tau} = \frac{1}{2\pi \times 0.002} = 79.6\ \mathrm{Hz} \tag{5.47}$$

We set the sample rate F_S to 50 times the bandwidth or 3980 Hz. The gain of the discrete system gain is computed using Equation 5.46:

$$G_D = \frac{1}{\tau F_S} = \frac{1}{0.002 \times 3980} = 0.126 \tag{5.48}$$

Finally, we adjust the gain value so that the product of all gains around the loop is 0.126:

$$G = \frac{0.126}{24} = 5.25 \times 10^{-3} \tag{5.49}$$

The system is shown in Figure 5.15 where the latch is represented in z-transform notation.[15]

Our strategy for comparing the response of the analog system with the discrete system is to use Equation 5.9 to compute the response of the analog

[15] A practical discussion of z-transforms is included in Chapter 8.

```
1     clear all;
2     close all;
3
4     Fs = 3980;  % Sample rate in Hz.
5
6     Ts = 1/Fs;  % Sample period in seconds.
7     NumSamples = round(6 * 0.002 * Fs);    % 6 time constants for display.
8     Tau = 0.002;
9     G = 1/(Tau*Fs);
10
11    Reg = 0;    % Digital register shown by z^-1
12    Output = zeros(1,NumSamples); % Storage for system output
13    Setpoint = 1;
14
15    for n = 1 : NumSamples
16        Error = Setpoint - Reg;
17        Reg = Reg + G*Error;
18        Output(1,n) = Reg;
19    end
20
21    % Compute analog system response.
22    n = 1 : NumSamples;
23    VoAnalog = 1.-exp(-n*Ts/Tau);    % Equation 49.
24
25    % Plot analog and discrete system response.
26    figure(1);
27    plot(n*Ts/.001,Output(1,:),'-k', 'LineWidth', 2);
28    hold on;
29    plot(n*Ts/.001,VoAnalog,':k',  'LineWidth', 2);
30    grid on;
31    xlabel('Time (ms)');
32    ylabel('Response (V)');
33    xlim([0 12]);
34    ylim([0 1.2]);
```

Figure 5.16. Matlab code used to compare analog and discrete control systems for Example 5.5.

controller and then use Matlab[16] to compute the response of the discrete controller and plot both responses. The response of an analog controller using the same time constant is

$$V_O(t) = 1\left(1 - e^{-t/0.002}\right) \tag{5.50}$$

The Matlab code shown in Figure 5.16 computes and plots the analog and discrete controller system response.

Figure 5.17 shows that the response of the discrete system closely matches the analog system.

[16]Microsoft Excel could be used for this also.

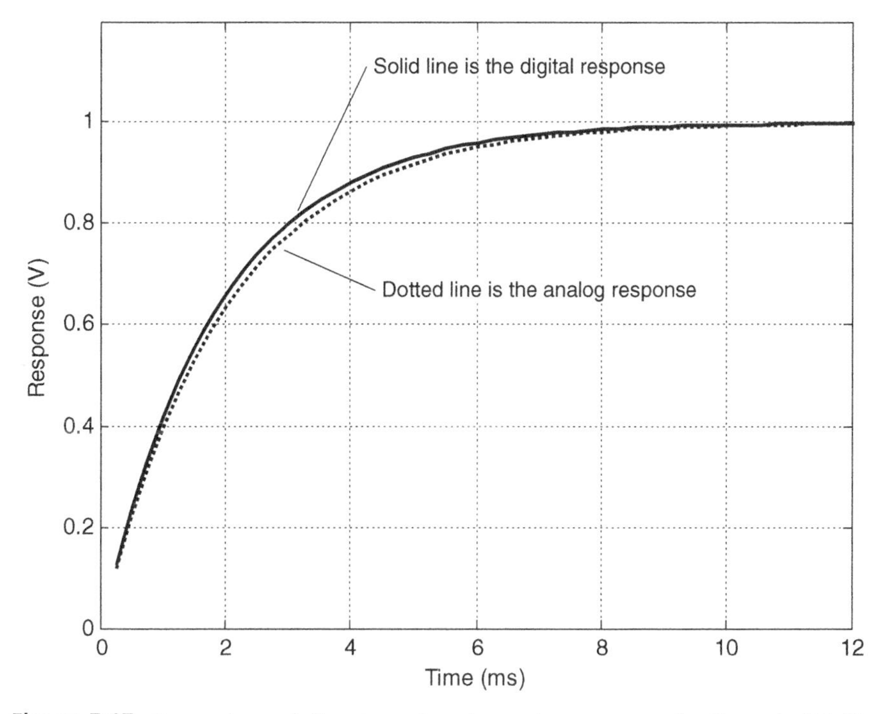

Figure 5.17. Comparison of discrete and analog system response for Example 5.5. The discrete sample rate is 50 times the bandwidth of the system.

PROBLEMS

5.1 Write an equation similar to Equation 5.1 that describes the system in Figure 5.18. Solve for the output $V_O(s)$ as a function of the input $V_{IN}(s)$ and the disturbance $D(s)$.

5.2 Say the system of Figure 5.5 has $G_A = 5$ and $H = 0.71$. Determine the required setpoint to make the steady-state system output equal to 1. With the steady-state system output equal to 1, what is the integrator voltage? Determine G_C so the time constant of the closed-loop system is 400 ms.

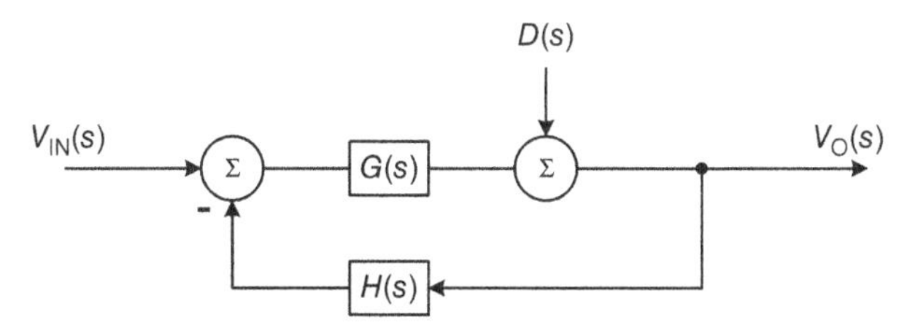

Figure 5.18. System for Problem 5.1.

5.3 Say the time constant of the actuator τ_1 in Figure 5.8 is 150 μs. Determine τ_2 so the damping ratio is $\xi = 0.4$. Determine the settling time and peak overshoot.

5.4 Show how to double the loop gain in the second-order circuit realization of Figure 5.12. Build or simulate the modified circuit and show that it matches the results of Figure 5.11.

5.5 A discrete time system using the topology of Figure 5.15 requires a time constant of 100 μs. Determine the minimum sample rate using the guideline in the chapter. Determine the gain. Show that the step response matches the response for the analog system with the same time constant using Microsoft Excel or a C program as your simulator.

REFERENCES

1. R. C. Dorf, *Modern Control Systems*, Prentice Hall, 2011.
2. Digi-Key Electronics—Electronics Components Distributor. *Digi-Key*. Available: www.digikey.com (Accessed 12/1/2013.)

6

HOW TO WORK WITH OP-AMP CIRCUITS

Operational amplifiers, or op-amps, satisfy many analog signal processing tasks including amplifying, level-shifting, and filtering. Op-amps are a workhorse of analog design below 1 MHz and are often used at higher frequencies. Knowing how to design basic op-amp circuits and knowing how to work with practical or nonideal op-amps are prime interview topics and key skills for entry-level engineers.

The op-amp has been used since the mid 1940s and is an excellent example of simplicity and cleverness; simple because it is just a high-gain differential amplifier and clever because of the many useful functions that can be implemented by surrounding it with a few external parts. In the 1950s, the author's father built op-amp circuits about the size of a brick using vacuum tubes. Later, he built op-amps out of handfuls of discrete transistors and he certainly appreciated the arrival of integrated circuit op-amps in the 1960s. Today's op-amps are small integrated circuits that use minimal power and can be purchased for less than a dollar. As op-amps continue to improve, they will satisfy an ever-increasing number of design applications, and one day as

Ten Essential Skills for Electrical Engineers, First Edition. Barry L. Dorr.

you look back at the improvements in op-amps in your career you'll probably be as impressed as the author's father.

As a circuit designer you'll certainly appreciate that when you properly design an op-amp circuit it will likely work as expected and require minimal troubleshooting. As your career progresses, you will encounter many clever and useful op-amp circuits and include them in your design toolbox. Hopefully you will have the satisfaction of inventing a few circuit topologies yourself.

This chapter begins by describing the op-amp as an ideal circuit element. This allows us to review commonly encountered circuit topologies such as the inverting and noninverting configurations. We then consider the characteristics of practical or nonideal op-amps. Understanding practical op-amps will enable you to select the best op-amps for your designs. Finally, we discuss the op-amp as a control system using the techniques covered in Chapter 5. This discussion will reinforce your knowledge of both op-amps and control systems.

6.1 THE IDEAL OP-AMP

The ideal op-amp is a simple circuit element that can be described in a few sentences. A real op-amp is usually described by a 50- to 100-page datasheet. We begin with the ideal op-amp because it is sufficient for many analysis and design steps.

The op-amp in Figure 6.1 is ideal because of the following:

1. The gain, defined as the output voltage divided by the voltage difference between the input terminals, or $V^+ - V^-$, is very high (~10,000) and considered infinite.

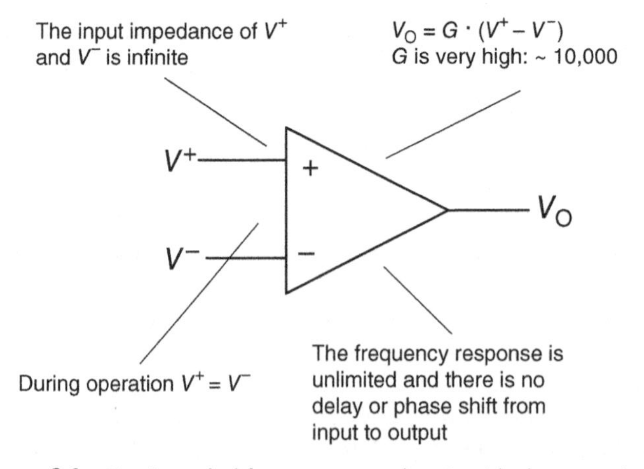

Figure 6.1. Circuit symbol for an op-amp showing ideal assumptions.

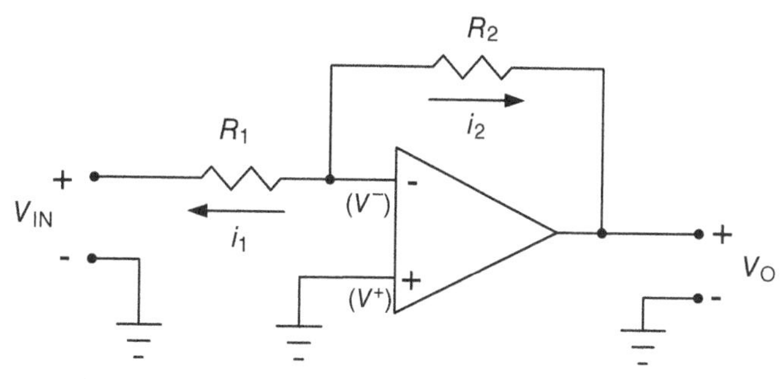

Figure 6.2. Inverting op-amp circuit. $V_O/V_{IN} = -R_2/R_1$.

2. The input impedance of terminals V^+ and V^- is infinite, meaning the input currents are zero.
3. The frequency response is unlimited. There is no delay or phase shift from input to output.
4. When the op-amp is used in a circuit with feedback, $V^+ = V^-$.[1]

The ideal assumptions above can be used to analyze basic op-amp circuits as shown in the following examples.

Example 6.1. Determine the voltage gain, V_O/V_{IN} for the op-amp circuit shown in Figure 6.2.

Solution. From the fourth assumption about ideal op-amps, we know that when the op-amp is used in a circuit with feedback, the voltages at V^+ and V^- are the same. Therefore the voltage at inverting terminal V^- is zero. From assumption 2, we know that there is no current flowing in or out of the inverting terminal. Applying Kirchhoff's current law at terminal V^- gives

$$i_1 + i_2 = 0 \tag{6.1}$$

where i_1 and i_2 are

$$i_1 = \frac{0 - V_{IN}}{R_1},\; i_2 = \frac{0 - V_O}{R_2} \tag{6.2}$$

[1]Note that this comes from the first assumption. Since the gain is very high, the difference between the V^+ and V^- terminals is essentially zero when the output is within the range of the supply voltage. Therefore, we use the approximation $V^+ = V^-$.

Substituting Equation 6.2 into Equation 6.1 gives the final answer

$$\text{Voltage gain} = \frac{V_O}{V_{IN}} = -\frac{R_2}{R_1} \tag{6.3}$$

This basic configuration is called an *inverting* amplifier because a positive input voltage produces a negative output and vice versa. This is why there is a negative sign in the gain formula.

The result in Equation 6.3 can be made even more useful by noting that R_1 and R_2 can be replaced by impedances Z_1 and Z_2 as shown in Equation 6.4. This allows substitution of reactive elements for the resistors, and the formula for gain $H(s)$ becomes a frequency-dependent transfer function. Using reactive elements allows op-amps to be used for active filtering applications as described in Chapter 7:

$$\text{Voltage gain} = \frac{V_O}{V_{IN}} = H(s) = -\frac{Z_2}{Z_1} \tag{6.4}$$

Example 6.2. Determine the transfer function of the circuit shown in Figure 6.3. Comment on what this circuit does.

Solution. The impedances of the resistor and capacitor are substituted into Equation 6.4. The transfer function is

$$\frac{V_O}{V_{IN}} = H(s) = -\frac{Z_2}{Z_1} = -\frac{\frac{1}{s}C}{R} = -\frac{1}{RC} \cdot \frac{1}{s} \tag{6.5}$$

Operation of the circuit can be recognized by noting that for a capacitor $i = C\,\mathrm{d}V/\mathrm{d}t$, meaning the voltage rises linearly with time when the current is constant. Say the capacitor is initially discharged and a voltage step is applied at the input. Since the op-amp will force the voltage at the inverting terminal

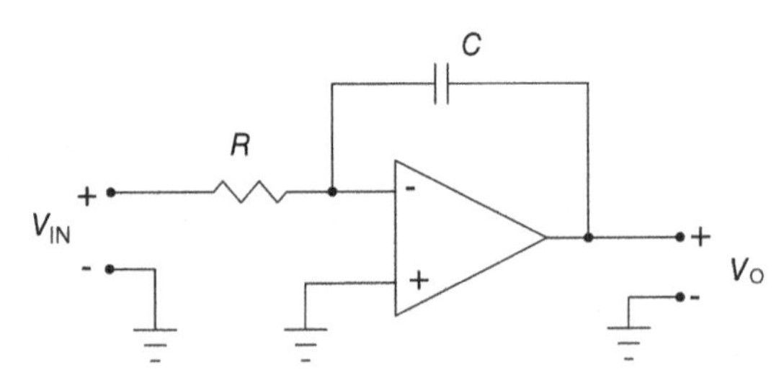

Figure 6.3. Integrating op-amp configuration. $V_O/V_{IN} = -1/RC \times 1/s$.

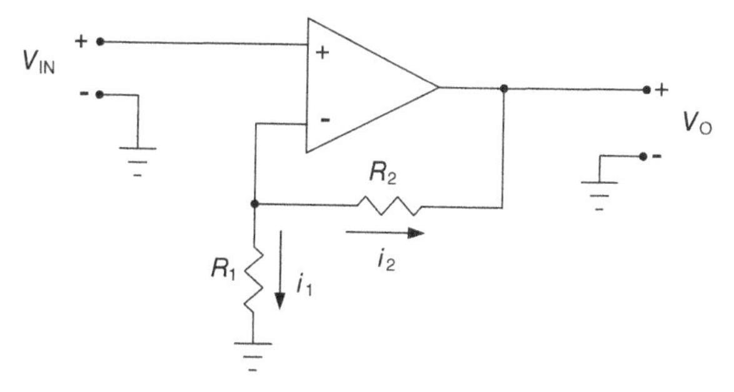

Figure 6.4. Noninverting op-amp circuit. $V_O/V_{IN} = 1 + R_2/R_1$.

to zero, the current in the resistor is constant. Since the current in the capacitor is the same as the current in the resistor, the voltage at the op-amp output must rise linearly. The current in the resistor is V_{Step}/R so the rate of change at the op-amp output can be computed as

$$\frac{dV_O}{dT} = \frac{i}{C} = \frac{V_{Step}}{RC} \tag{6.6}$$

The analysis above shows that if a voltage step is applied, the output of the circuit increases linearly with slope equal to the step voltage divided by the time constant.[2] Operation of the circuit could also be determined by noting that the $1/s$ in the transfer function is the Laplace transform of an integrator. Integrating op-amp circuits are commonly found in control systems as described in Chapter 5.

Example 6.3. Determine the voltage gain V_O/V_{IN} for the op-amp circuit shown in Figure 6.4.

Solution. Since the voltages at the input terminals must be the same, the voltage at the inverting terminal is V_{IN}. Applying Kirchhoff's current law at the inverting terminal gives

$$\frac{V_{IN} - V_O}{R_2} + \frac{V_{IN}}{R_1} = 0 \tag{6.7}$$

[2]This description shows how to determine the operation of a circuit using just a few basics. Students are frequently hesitant to attempt this on a job interview—even if the interviewer 'prods' them through the process. The reader is encouraged to practice this sort of analysis with any circuit that is not immediately recognizable. With practice you will get surprisingly skilled at quickly analyzing circuits.

Solving Equation 6.7 for the voltage gain gives

$$\text{Voltage gain} = \frac{V_O}{V_{IN}} = 1 + \frac{R_2}{R_1} \tag{6.8}$$

6.2 PRACTICAL OP-AMPS

The assumption of ideality is a powerful tool because it allows the op-amp to be used as a simple circuit element, thereby allowing the designer to focus on the "big picture" of a design or analysis task. However, there are practical or nonideal considerations that must be taken into account for a design to work as expected.

The datasheet for the Texas Instruments TLC 272 op-amp [1] contains a wealth of useful design information. The remainder of this chapter focuses on several key specifications and concepts which will help you use op-amps effectively.

6.2.1 Effect of Input Offset Voltage

One of the ideal assumptions is that the voltage across the input terminals of the op-amp is zero. Real op-amps have a small imbalance in their input stage and a voltage called the input offset voltage is required to drive the output voltage to zero. Figure 6.5 shows an ideal op-amp with the input offset voltage represented by a voltage source. Typical input offset voltages are on the order of ±3 to ±10 mV and vary from part to part and with temperature.

Input offset voltage is a critical specification because its effect on the output is based on the gain of the circuit. Consider the effect of input offset voltage on the output of the inverting amplifier shown in Figure 6.6.

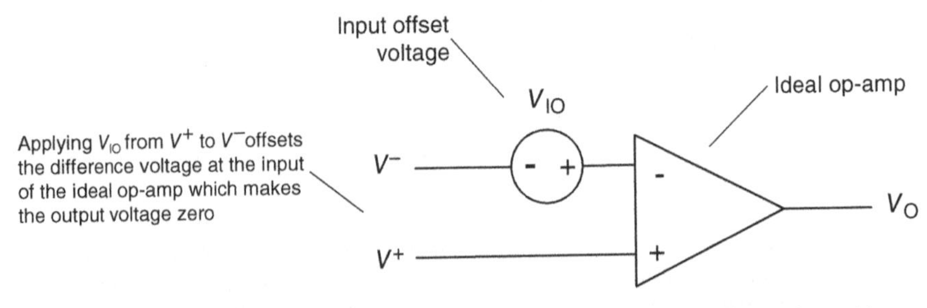

Figure 6.5. Ideal op-amp modified to show the effect of input offset voltage V_{IO}.

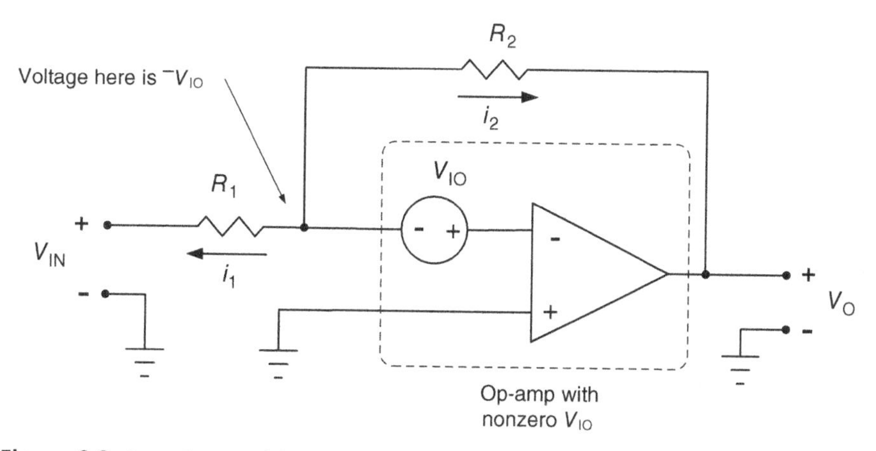

Figure 6.6. Inverting amplifier with nonzero input offset voltage. V_{IO} is critical because the output voltage is V_{IO} multiplied by $1 + R_2/R_1$.

The analysis is similar to Example 6.1. Maintaining the ideal assumption of no current flowing in or out of the input terminals, Kirchhoff's current law is applied at the junction of R_1 and R_2:

$$\frac{-V_{IO} - V_O}{R_2} + \frac{-V_{IO} - V_{IN}}{R_1} = 0 \tag{6.9}$$

Rearranging to solve for V_O gives

$$V_O = -V_{IN}\frac{R_2}{R_1} - V_{IO}\left(1 + \frac{R_2}{R_1}\right) \tag{6.10}$$

Equation 6.10 shows that the offset at the circuit output is the input offset voltage multiplied by $-\left(1 + R_2/R_1\right)$. As the circuit gain, $-R_2/R_1$, increases, so does the magnitude of the DC offset at the circuit output. This is why input offset voltage is normally shown on the first page of most op-amp datasheets.

Example 6.4. Figure 6.7 shows a sensor[3] connected to an inverting amplifier. The sensor outputs a DC voltage between 0 and –33 mVDC. The accuracy of the sensor is ±3%. The amplifier multiplies the sensor output by –100 which sets the range of the op-amp output from 0 to 3.3 V. The input offset voltage

[3]The sensor might measure temperature, position, speed, and so on.

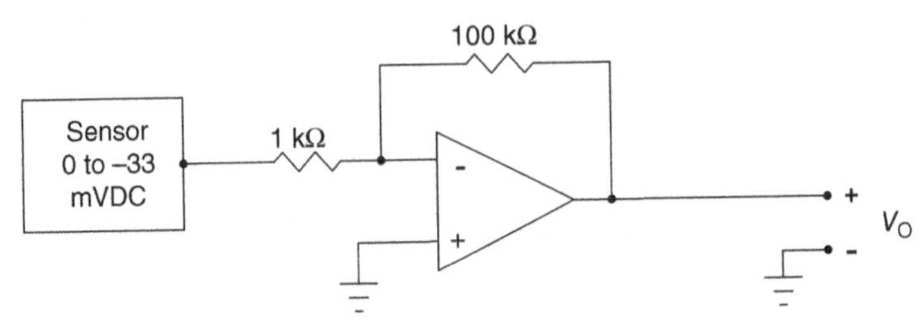

Figure 6.7. Circuit for Example 6.4. The input offset voltage of the op-amp degrades the accuracy of this measurement system.

specification for the op-amp is ±3 mV. Determine the range of DC offset at the op-amp output and comment on the usefulness of this design.

Solution. From Equation 6.10 we see that the output voltage due to V_{IO} is

$$V_O = -V_{IO}\left(1 + \frac{100{,}000}{1000}\right) = \pm 3\,\text{mV} \times -101 = \pm 303\,\text{mV} \quad (6.11)$$

The op-amp causes a ±303 mV error in an output intended to range from 0 to 3.3 V which is about ±9%. Our sensor has an accuracy specification of ±3% so we conclude that our circuit degrades the accuracy of this system.[4] This example highlights what can happen if an op-amp is used without taking its specifications into account.

6.2.2 Noise Contribution from Op-Amp Circuits

When the input of an op-amp circuit is grounded, a small amount of noise can be seen at its output that can degrade performance in some designs. Noise is generated inside the op-amp and by the resistors used to set the gain. A full discussion of op-amp noise is beyond the scope of this book, but is well covered in [2]. Since noise generated by the op-amp itself is generally the largest noise contributor, it is the focus of this section.

Figure 6.8 shows the noise specification, given as a function of frequency, for the TLC272 op-amp. At frequencies below about 100 Hz, the noise specification is dominated by flicker noise which decreases as the frequency increases. Above 1 kHz the noise specification is dominated by thermal noise

[4] Accurate sensors are generally expensive. If we pay for an accurate sensor we should take full advantage of its accuracy.

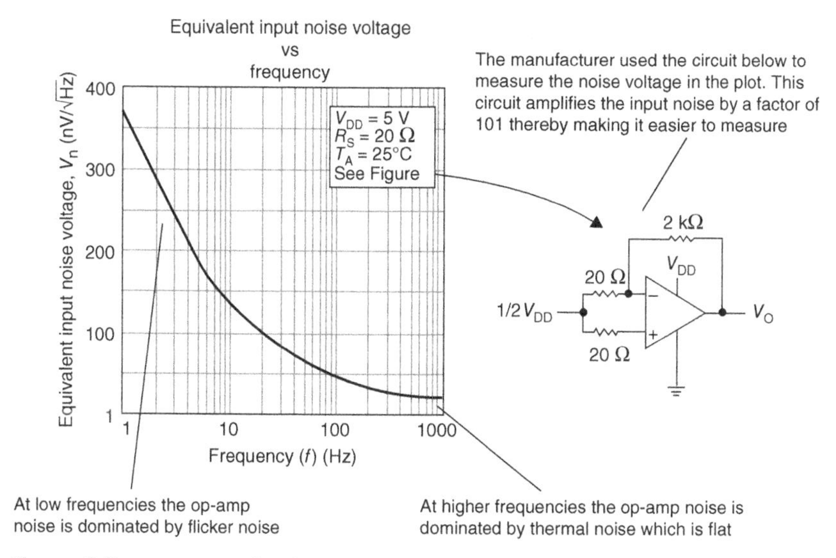

Figure 6.8. Input noise for the TLC272 op-amp. (Plot courtesy of Texas Instruments)

which is constant or flat with frequency. The value in the flat part, $25\text{nV}/\sqrt{\text{Hz}}$, is the noise specification given in the TLC272 datasheet.

The noise specification is a density function. This means that to compute the total effective noise voltage at the input, the curve of Figure 6.8 must be integrated over the bandwidth of interest. It's actually slightly more difficult because the vertical axis of the plot has units of $\text{nV}/\sqrt{\text{Hz}}$ and the horizontal axis has units of Hz, which means we must square the integrand prior to integration and then take the square root of the result as shown in Equation 6.12:

$$V_N = \sqrt{\int_{f_l}^{f_h} n^2\, df} \tag{6.12}$$

where V_N is the root mean square (RMS) value of the noise voltage referred to the op-amp input.[5] f_l is the lower end of the bandwidth of interest in Hz. f_h is the upper end of the bandwidth of interest in Hz. n is the op-amp noise specification in the bandwidth of interest in $\text{nV}/\sqrt{\text{Hz}}$.

[5]Manufacturers specify noise at the op-amp input so the designer can use the specification for any gain value or topology.

Example 6.5. Compute the equivalent RMS input-referred noise for the TLC272 in the bandwidth from 3000 to 10,000 Hz.

Solution. The noise specification between 3000 and 10,000 Hz is $25\,\text{nV}/\sqrt{\text{Hz}}$ because this frequency range is in the flat part of Figure 6.8. We evaluate the integral of Equation 6.12:

$$\text{RMS input noise voltage} = V_{\text{RMS}} = \sqrt{\int_{3000}^{10{,}000} \left(25 \times 10^{-9}\right)^2 df} \tag{6.13}$$

$$V_{\text{RMS}} = \sqrt{(10{,}000 - 3000)\left(25 \times 10^{-9}\right)^2} = \sqrt{\text{BW}} \times 25 \times 10^{-9} = 2.09\ \mu\text{VRMS} \tag{6.14}$$

In the above example our bandwidth of interest is in the flat part of the op-amp's noise curve. When this is the case the noise can be computed by simply multiplying the value from the curve by the square root of the bandwidth. If the bandwidth of interest includes frequencies where flicker dominates, use the useful formula on page 15 of [2].

The noise voltage at the op-amp output depends on the gain of the circuit and the input-referred noise. Input noise is referred to the output in exactly the same way as input offset voltage. Therefore, as suggested by Figure 6.6 and Equation 6.10 the noise at the output of the inverting op-amp is

$$V_{\text{NO}} = -V_{\text{NI}}\left(1 + \frac{R_2}{R_1}\right) \tag{6.15}$$

where V_{NO} is the noise in $\text{V}/\sqrt{\text{Hz}}$ at the op-amp output and V_{NI} is the noise in $\text{V}/\sqrt{\text{Hz}}$ at the op-amp input.

Example 6.6. Compute the RMS noise voltage at the output of the op-amp shown in Figure 6.9. The op-amp's noise specification is $18\ \text{nV}/\sqrt{\text{Hz}}$ in the region dominated by thermal noise and the bandwidth of interest is 12–39 kHz.

Solution. The input-referred noise computed from Equation 6.12 (see Example 6.5) is

$$\text{Input-referred noise} = 18\frac{\text{nV}}{\sqrt{\text{Hz}}}\sqrt{39{,}000 - 12{,}000} \cdot \sqrt{\text{Hz}} = 2.96\ \mu\text{VRMS} \tag{6.16}$$

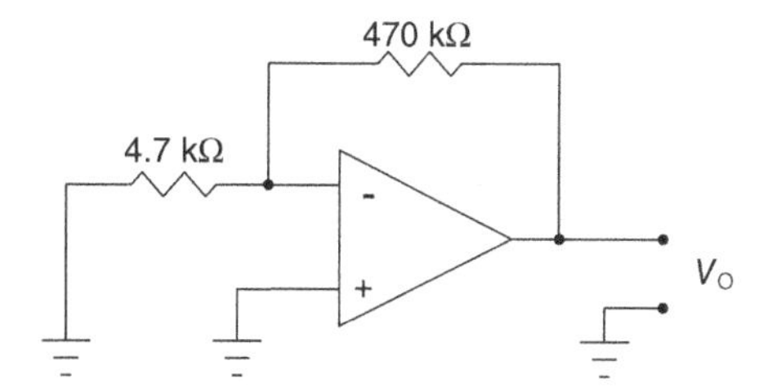

Figure 6.9. Circuit for Example 6.6. The noise voltage at the output of the op-amp is determined by the input-referred noise and the gain of the circuit.

And the output noise is computed using Equation 6.15:

$$\text{Output-referred noise} = 2.96 \times 10^{-6}\left(1 + \frac{470{,}000}{4700}\right) = 299\ \mu\text{VRMS} \tag{6.17}$$

6.2.3 Dynamic Characteristics of Op-Amp Circuits

One of the ideal assumptions earlier was that the op-amp has unlimited frequency response, no delay, and no phase shift. In other words, we assumed the relationship between the input and output is

$$V_O = G \cdot (V^+ - V^-) \tag{6.18}$$

where G is a very large constant.

In reality, the op-amp gain, G, decreases with frequency and its phase shift increases with frequency making G a frequency-dependent transfer function, $G(s)$, which is shown for the TLC272 op-amp in Figure 6.10. The effects of the frequency dependence are limited bandwidth for op-amp circuits, overshoot when a step input is applied, and possibly instability. This section focuses on instability because it can cause catastrophic circuit failures.

To evaluate stability, we must view the op-amp from a control systems perspective. This may seem difficult at first, but it will solidify your understanding of both op-amp circuits *and* control systems. We will also use this perspective to clear up several issues that tend to confuse students and experienced engineers alike.

Figure 6.11 shows how to view a noninverting op-amp circuit (a) as a classical control system. First the op-amp is represented as a summing junction followed by a high-gain amplifier and a lowpass filter (b). Next the amplifier and filter are combined into the block marked $G(s)$, and R_2 and R_1 are shown as feedback gain $H(s)$ in (c).

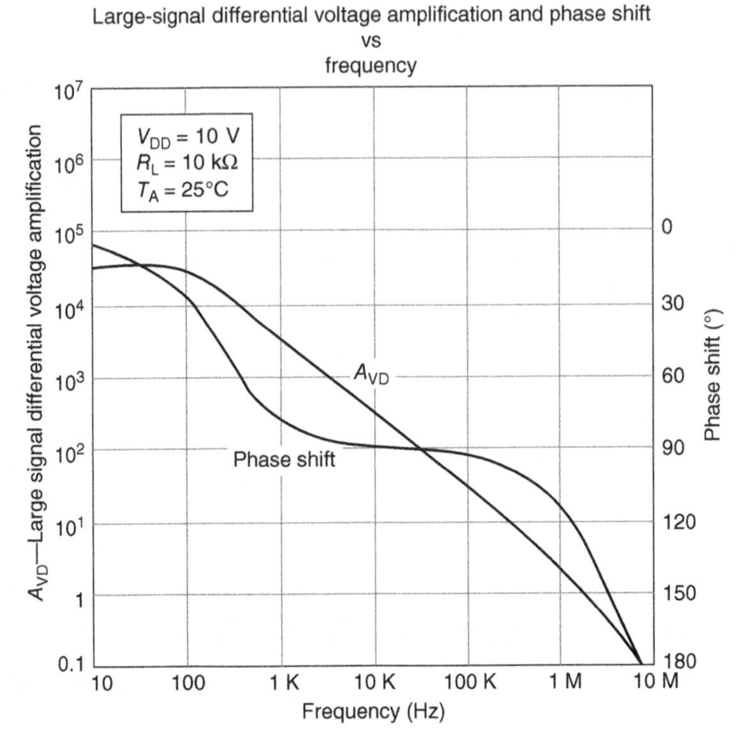

Figure 6.10. Gain and phase plot for TLC272 op-amp. (Plot courtesy of Texas Instruments)

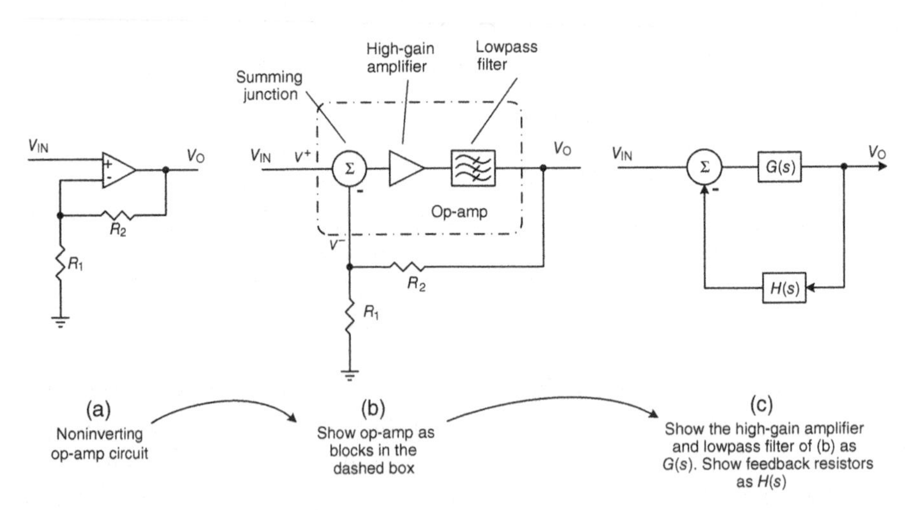

Figure 6.11. The steps in this figure show how a noninverting op-amp circuit can be represented as a classical control system as described in Chapter 5. The control perspective is useful for understanding dynamic characteristics of op-amp circuits.

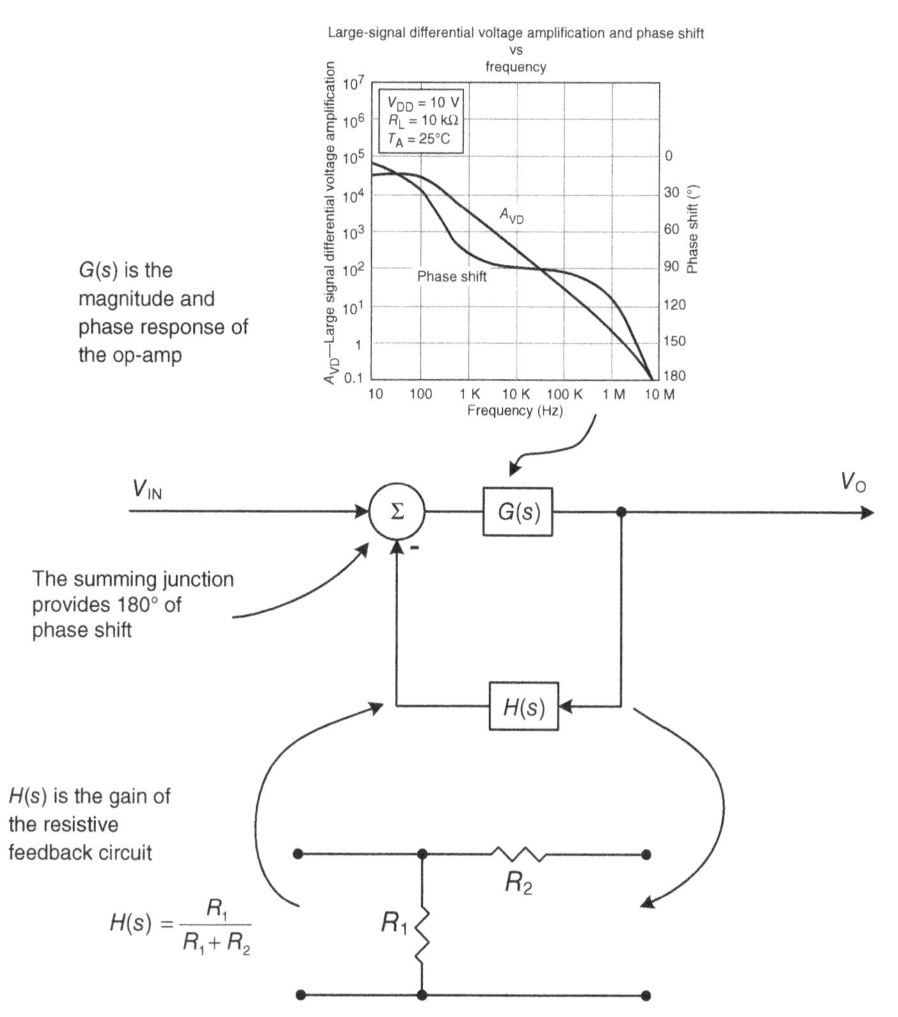

Figure 6.12. Control systems interpretation of the noninverting op-amp. $G(s)$ and $H(s)$ describe the forward and feedback paths.

The control systems viewpoint is emphasized in Figure 6.12. Treating the op-amp circuit as a feedback loop allows us to verify that it is *stable*, meaning that it will not oscillate. Oscillations occur when small perturbations within the feedback loop, such as those generated by thermal noise in the resistors, are amplified by the loop and the resulting signals add in phase to create large, undesired AC signals. A necessary, but not sufficient criteria for stability is that the product of all gains around the loop (the cumulative gain) must be less than unity at any frequency where the sum of all phase shifts (the cumulative phase) is a multiple of 360°. Referring to Figure 6.12, the summing junction

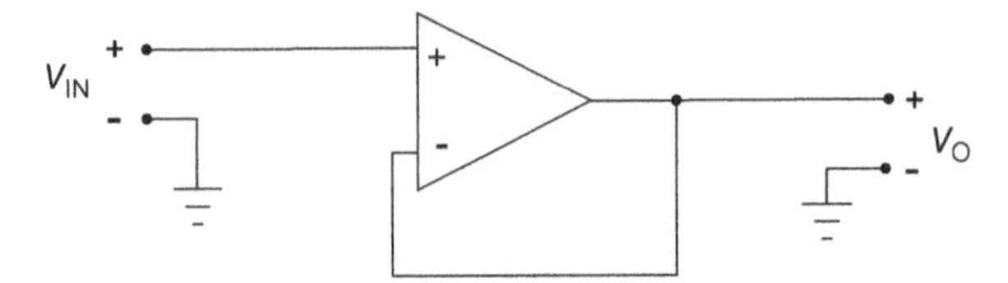

Figure 6.13. Voltage follower is equivalent to a noninverting amplifier with $R_2 = 0$ and $R_1 = \infty$. This configuration has a voltage gain of 1 and is used to isolate one section of a circuit from another.

provides 180° of phase shift, and the feedback resistors do not contribute to the phase,[6] and the system is stable if the magnitude of the product $G(s)H(s)$ is less than unity when the phase of $G(s)$ is equal to –180°.

The circuit shown in Figure 6.13 is called a *voltage follower* and is used in practice to isolate one section of a circuit from another. It is essentially a noninverting configuration with $R_2 = 0$ and $R_1 = \infty$, so Equation 6.8 shows its gain is unity.

For the voltage follower $H(s) = 1$, and the cumulative gain is the same as the gain represented by the line marked A_{VD} in Figure 6.14. The plot shows that at 180° phase shift the gain is just less than 0.1 so the voltage follower is stable. Now suppose feedback resistors $R_2 = 9\ \text{k}\Omega$ and $R_1 = 1\ \text{k}\Omega$ are added to the noninverting circuit of Figure 6.11a. The circuit gain from Equation 6.8 is now 10 and the gain of the feedback network is 0.1. This lowers the cumulative loop gain by a factor of 10 as shown by the dotted line in Figure 6.14. Since the gain at the 180° phase point is lower than it was in the voltage follower configuration, the circuit is more stable.

The previous comments point out an important concept that seems to confuse students and experienced engineers alike: We correctly associate high cumulative or *loop* gain with a tendency for a system to oscillate. But for op-amp circuits *high loop gain* results in *low circuit gain* and vice versa. Op-amp datasheets frequently state that the op-amp is "unity gain stable" to convey to the designer that the part is stable at the highest possible loop gain.

6.2.4 Effect of Capacitive Loading

Nearly all commercially available op-amps are unity gain stable. However, a capacitive load placed at the output of an op-amp has the effect of lowering the phase shift curve without affecting the gain as suggested in Figure 6.15. This causes the phase to pass through 180° at a lower frequency where the

[6]This is not *always* true. See Problem 6.6.

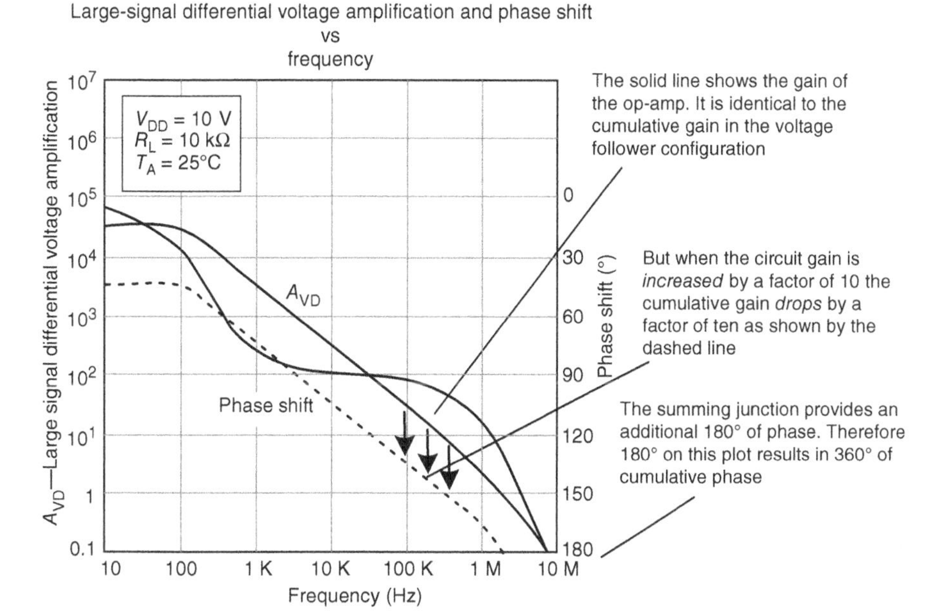

Figure 6.14. The open-loop plot can be used as a closed-loop plot by including the effect of the feedback resistors. This plot shows that *raising* the circuit gain *lowers* the cumulative gain thereby making the circuit more stable. (Plot courtesy of Texas Instruments)

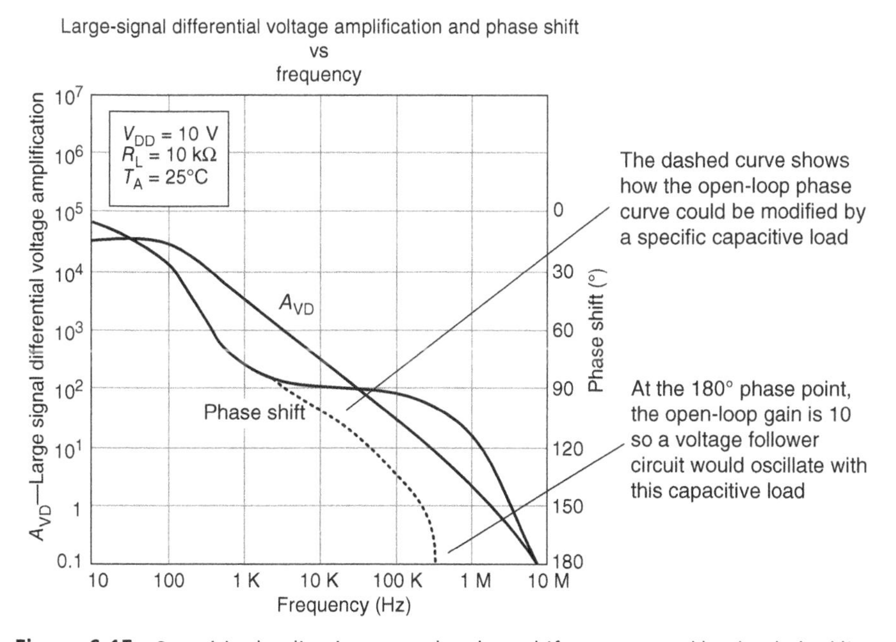

Figure 6.15. Capacitive loading increases the phase shift as suggested by the dashed line. In this example the op-amp with capacitive loading would oscillate if connected as a voltage follower. (Plot courtesy of Texas Instruments)

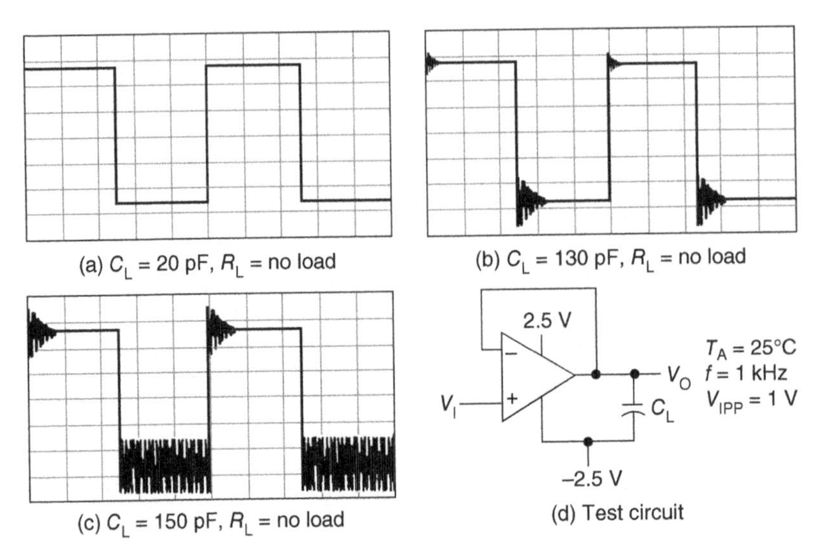

Figure 6.16. The TLC272 datasheet shows the effect of capacitively loading a voltage follower using the test circuit in (d). A 20 pF load (a) has no noticeable effect, but capacitive loads of 130 pF (b) and 150 pF (c) reduce stability and result in ringing. (Plot courtesy of Texas Instruments)

gain is higher, thereby making the circuit less stable and prone to overshoot in response to step inputs.

The datasheet for the TLC272 provides excellent data for capacitive loading as well as simple circuit corrections to accommodate it. Figure 6.16 shows the effect of capacitively loading a voltage follower circuit when a step input is applied.

6.2.5 A Nagging Issue

What follows is an issue with op-amp circuits that stumps many students and experienced hardware engineers. From the previous discussion and your knowledge of control systems you'll grasp it immediately and gain a reinforced understanding of op-amp circuits. The question is "Given that the open-loop plot of Figure 6.10 shows 30° phase shift at 100 Hz, why does a closed-loop op-amp circuit have essentially no phase shift at 100 Hz?" The answer is found by viewing the op-amp circuit from the control systems perspective.

Rather than try to remember the transfer function for a closed-loop control system we derive it. See Chapter 5. Inspection of Figure 6.17 gives the equation below:

$$V_O(s) = \left[V_{IN}(s) - V_O(s) \cdot H(s)\right] G(s) \tag{6.19}$$

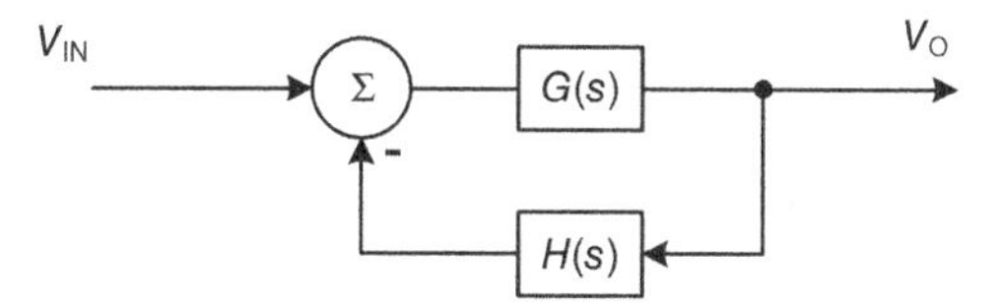

Figure 6.17. Control system block diagram.

which reduces to

$$\frac{V_O(s)}{V_{IN}(s)} = \frac{G(s)}{1 + G(s)H(s)} \tag{6.20}$$

From our discussion in Section 6.2.3 we know that the circuit feedback, $H(s)$, is a resistive network and its transfer function has no frequency dependence, so we drop its dependence on s. We also see from Figure 6.10 that the magnitude of the op-amp gain, $G(s)$, at 100 Hz is about 30,000. Therefore Equation 6.20 becomes

$$\frac{V_O(s)}{V_{IN}(s)} \cong \frac{G(s)}{G(s) \cdot H} = \frac{1}{H} \tag{6.21}$$

Equation 6.21 shows that if the magnitude of the open-loop gain is large, then $G(s)$, which contains the phase of the op-amp, cancels in the closed-loop gain expression. Therefore at 100 Hz, the transfer function of the op-amp circuit has essentially no phase shift.

PROBLEMS

6.1 Determine V_O as a function of V_A, V_B, R_2, and R_1 for the circuit in Figure 6.18. *Hint:* Use superposition, that is, first compute the output due to V_A with V_B grounded, and then compute the output due to V_B with V_A grounded. Sum the responses to get the final result.

6.2 Determine V_O for the circuit in Figure 6.19. Substitute $V_{IN} = s(t) + V_{Bias}$ to show that the circuit amplifies the signal but passes the bias unamplified. *Hint:* Use superposition to solve this problem.

6.3 The dotted line in Figure 6.15 shows the effect on the open-loop phase shift due to capacitive loading of a voltage follower. Determine how much the *loop* gain must be reduced to make the circuit stable. What is the gain for the resulting noninverting amplifier?

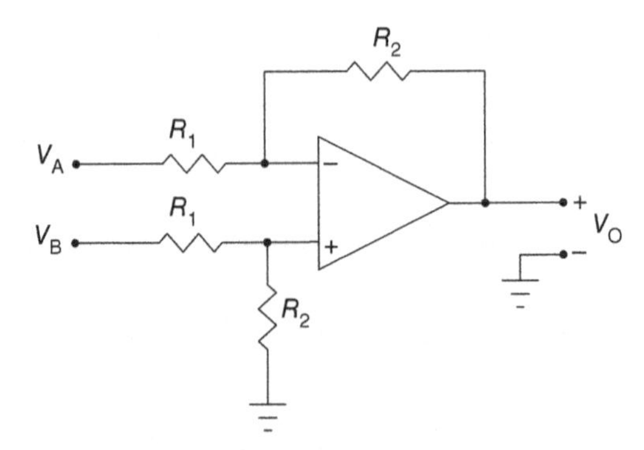

Figure 6.18. Circuit for Problem 6.1. This circuit is called a *difference amplifier.*

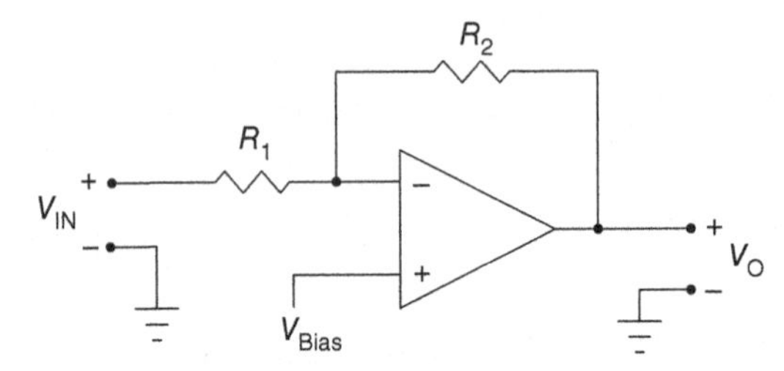

Figure 6.19. Circuit for Problem 6.2. This configuration is commonly used when only a positive power supply is available.

6.4 Show that the magnitude of the effect of input-referred noise and V_{IO} on the noninverting amplifier is the same as for the inverting amplifier.

6.5 A sensor is connected to an op-amp as shown in Figure 6.20. The op-amp output connects to a 10-bit ADC with full range voltage of 0 –1.8 V. Determine the maximum V_{IO} so that the amplifier offset error will cause less than 5 ADC quanta.

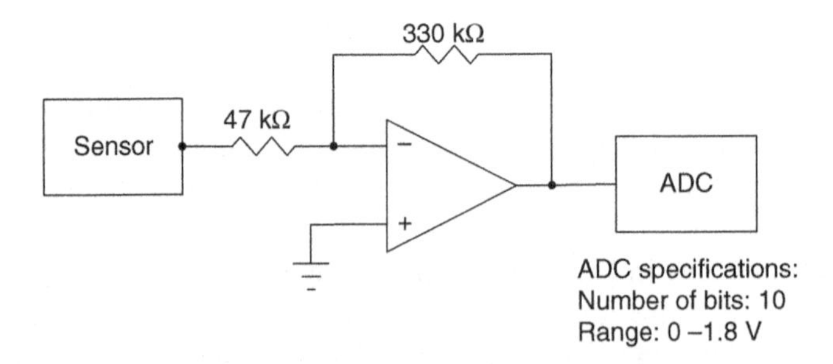

Figure 6.20. Circuit for Problem 6.5.

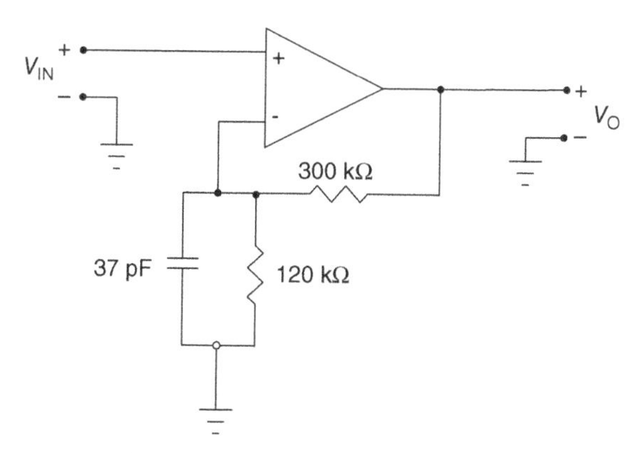

Figure 6.21. Circuit for Problem 6.6.

6.6 The circuit in Figure 6.21 has 37 pF stray capacitance from the inverting terminal to ground. Determine the transfer function $V_O(s)/V_{IN}(s)$ for the circuit taking into account the stray capacitance. Determine R_1 and R_2 so the DC gain of the circuit is unchanged, but the capacitance affects the gain by no more than 3 dB at 100 kHz.

6.7 A TLC272 op-amp is used in a noninverting configuration with a gain of 5. Figure 6.10 shows that the open-loop gain of the TLC272 op-amp at 300 kHz is 10 and the phase is –105°. Determine the magnitude and phase of the noninverting amplifier at 300 kHz. *Hint:* See Section 6.2.5.

REFERENCES

1. Texas Instruments. *TLC272 Operational Amplifier Data Sheet*, 2002. Available: http://www.ti.com/lit/ds/symlink/tlc272b.pdf (Accessed 11-30-2013.)
2. Texas Instruments. *Noise Analysis in Operational Amplifier Circuits*, 2007. Available: http://www.ti.com/lit/an/slva043b/slva043b.pdf (Accessed 11-30-2013.)

7

HOW TO DESIGN ANALOG FILTERS

Filters are used to pass or reject signals based on their frequency content. Filter applications vary widely; a radio receiver uses filtering to isolate a single desired station from all of the signals arriving at its antenna, a radio transmitter uses filtering so that it emits signals only in the frequency band for which it is licensed, a measurement system uses filters to remove noise from sensor inputs, and printed circuit boards use bypass capacitors to filter the power connections to integrated circuits. Analog filters can be implemented with active topologies using op-amps, or passive topologies using inductors and capacitors.

Once a filter is fully specified, the components can easily be determined using excellent textbooks and online resources. The skill of specifying the filter is more valuable than the skill of selecting the components because it requires knowledge of the system design goals, knowledge of the signals to be filtered, knowledge of the components and systems surrounding the filter, and knowledge of the different filter types. The examples and problems in this chapter are all designed to give you experience converting system specifications into filter specifications and ultimately into practical filters.

Ten Essential Skills for Electrical Engineers, First Edition. Barry L. Dorr.

This chapter begins with a discussion of passive versus active filters. This is followed by a detailed discussion of the commonly encountered lowpass RC filter. Next we describe characteristics of three common filter types: Butterworth, Chebyshev, and Bessel. Understanding these characteristics will help you select the best filter type for a design requirement. Section 7.1 describes how filters are designed based on normalized lowpass prototypes, and Sections 7.6 and 7.7 show practical examples of active and passive filter design.

7.1 PASSIVE VERSUS ACTIVE FILTERS

Analog filters are classified as either passive or active. The major difference between them is that the active filter shown in Figure 7.1a uses the op-amp as a source of energy whereas the passive filter in Figure 7.1b has no internal energy source.

Active filters consist of resistors, capacitors, and active elements that are typically op-amps. An advantage of active filters is that they tend to be physically smaller and less expensive than passive filters because they do not use inductors. Another advantage is that active filters are generally unaffected by source and load impedances. A disadvantage is that the op-amps and the resistors generate noise as shown in Chapter 6 which makes these filters unusable for amplifying low-level signals such as those picked up by an antenna in a radio receiver. Another disadvantage is that, unlike passive filters, active filters cannot be used to filter high-power signals such as those generated by a radio transmitter. Finally, the frequency range of active filters is limited because they require op-amps with gain-bandwidth products on the

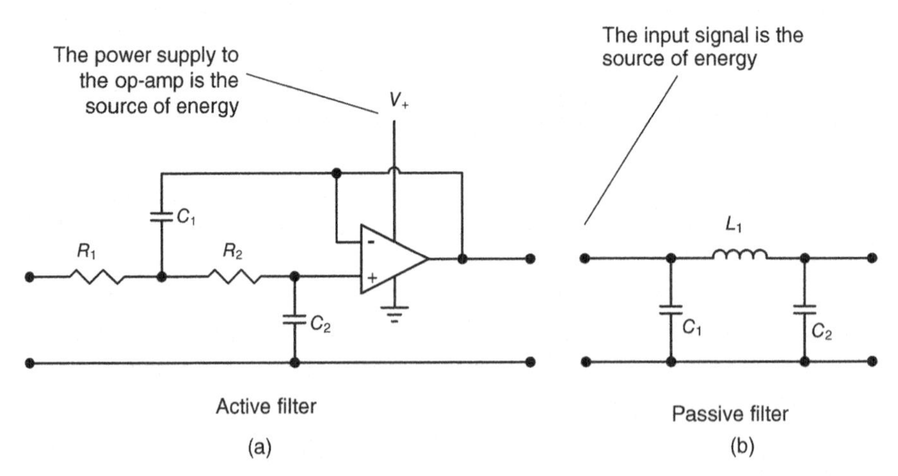

Figure 7.1. Active and passive filters. The active filter (a) has an energy source whereas the passive filter (b) does not.

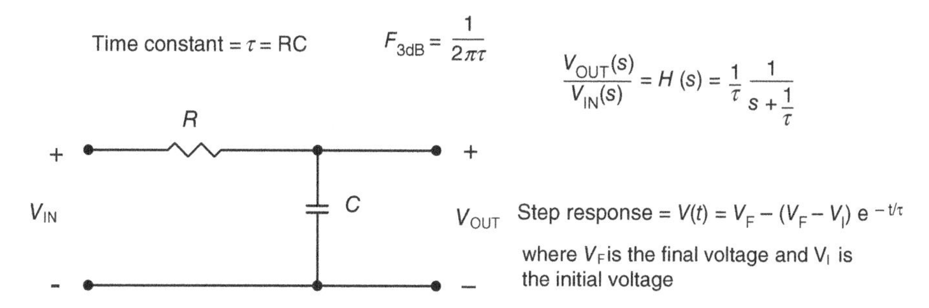

Figure 7.2. The lowpass RC filter is simple to design and implement using the equations shown here.

order of 10 or 20 times the filter bandwidth. However, a wide variety of filter designs are not affected by these disadvantages and active filters are nearly always used at frequencies less than 10 MHz and often at higher frequencies.

Passive filters are used when active filters are not a practical choice. They consist of capacitors and inductors, and the inductors make them physically larger and costlier than active filters. Passive filters generate no noise and can filter the femtowatts entering a radio receiver. They can also be used to filter high-power signals. And, of course, they don't require a power supply.

7.2 THE LOWPASS RC FILTER

The lowpass RC filter, shown in Figure 7.2, is typically used when the signals to be removed are at least several octaves higher in frequency than the desired signal. This is common when high-speed digital signals are located near low-frequency analog signals. Since the components are inexpensive, designers frequently include this simple filter on analog signal traces even if they are not sure filtering will be needed.[1]

The transfer function relating the output voltage to the input voltage can be written using the voltage divider theorem:

$$\frac{V_O}{V_{IN}} = \frac{\frac{1}{sC}}{R + \frac{1}{sC}} = \frac{1}{RC} \cdot \frac{1}{s + \frac{1}{RC}} = \frac{1}{\tau} \cdot \frac{1}{s + \frac{1}{\tau}} \tag{7.1}$$

where $\tau = RC$ is the time constant for the circuit.

[1]This practice is common in industry but should be done carefully. If the signal trace is part of a feedback loop, the phase shift from the *RC* filter could potentially cause instability. See Chapter 5.

Letting $s = j2\pi f$, and taking the absolute value, the magnitude of the frequency response is

$$\left|\frac{V_O}{V_{IN}}\right| = \frac{1}{\tau}\frac{1}{\sqrt{\frac{1}{\tau^2} + (2\pi f)^2}} = \frac{1}{\sqrt{1 + (2\pi f\tau)^2}} \tag{7.2}$$

Equation 7.2 can be expressed in decibels as

$$\text{Gain}_{\text{dB}} = 20\log_{10}\left(\frac{1}{\sqrt{1 + (2\pi f\tau)^2}}\right) \tag{7.3}$$

A common figure of merit for any filter is the 3 dB or cutoff frequency.[2] For the lowpass RC filter this is the frequency where

$$\left|\frac{V_O}{V_{IN}}\right| = \frac{1}{\sqrt{2}} = 0.7071 \tag{7.4}$$

The 3 dB frequency is computed by solving Equation 7.2 for f and setting $|V_O/V_{IN}|$ to 0.7071:

$$F_{3\text{dB}} = \frac{1}{2\pi\tau} = \frac{1}{2\pi RC} \tag{7.5}$$

Equation 7.2 is exact, but the Bode plot shown in Figure 7.3 gives a quick, accurate approximation of the magnitude of the frequency response. It is plotted by approximating the response as flat from DC to the 3 dB frequency, and then falling to –6 dB at twice the 3 dB frequency, –12 dB at four times the 3 dB frequency, and so on.

We can use the notion of the Bode plot to quickly approximate the attenuation[3] of the lowpass RC filter at any frequency. We start by expressing the ratio of an arbitrary frequency, F, to the 3 dB frequency in octaves:

$$2^n = \frac{F}{F_{3\text{dB}}} \tag{7.6}$$

[2]These terms "3 dB frequency" and "cutoff frequency" are used interchangeably in this chapter.

[3]Attenuation is the inverse of gain. When expressed in decibels, $\text{Atten}_{\text{dB}} = -\text{Gain}_{\text{dB}}$.

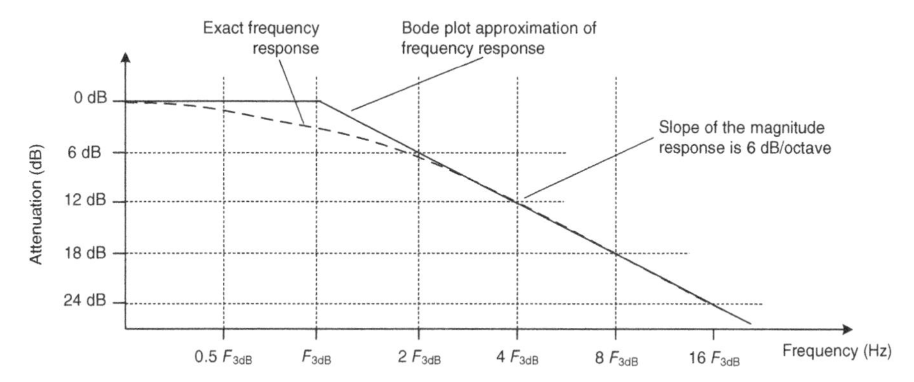

Figure 7.3. Frequency response plot for the lowpass RC filter. The Bode approximation (solid line) is an accurate approximation of the exact frequency response (dashed line) at low and high frequencies.

where n is the number of octaves.[4] Solving for n gives the useful formula

$$\text{Number of octaves between } F \text{ and } F_{3\text{dB}} = n = \frac{\ln\left(\frac{F}{F_{3\text{dB}}}\right)}{\ln(2)} \tag{7.7}$$

Since the lowpass RC filter rolls off at 6 dB per octave, the attenuation at any frequency an octave or more above the 3 dB frequency can be quickly determined using Equation 7.7 as shown in the following example.

Example 7.1. Design a filter for a data acquisition system that acquires signals from DC to 100 Hz. The desired signal is corrupted by an undesired digital clock signal at 10 kHz. The filter should attenuate the desired signal less than 3 dB and the undesired signal by at least 35 dB.

Solution. We can quickly determine if a lowpass RC filter will suffice. The number of octaves between 10 kHz and 100 Hz is given by Equation 7.7:

$$\text{Octaves} = \frac{\ln\left(\frac{10{,}000}{100}\right)}{\ln(2)} = 6.644 \tag{7.8}$$

Multiplying the number of octaves by 6 dB/octave gives the attenuation at 10 kHz as 39.9 dB. If the 3 dB bandwidth were set to 100 Hz, the desired signal would be attenuated by 3 dB and the undesired signal would be attenuated by

[4] An *octave* is a doubling of frequency.

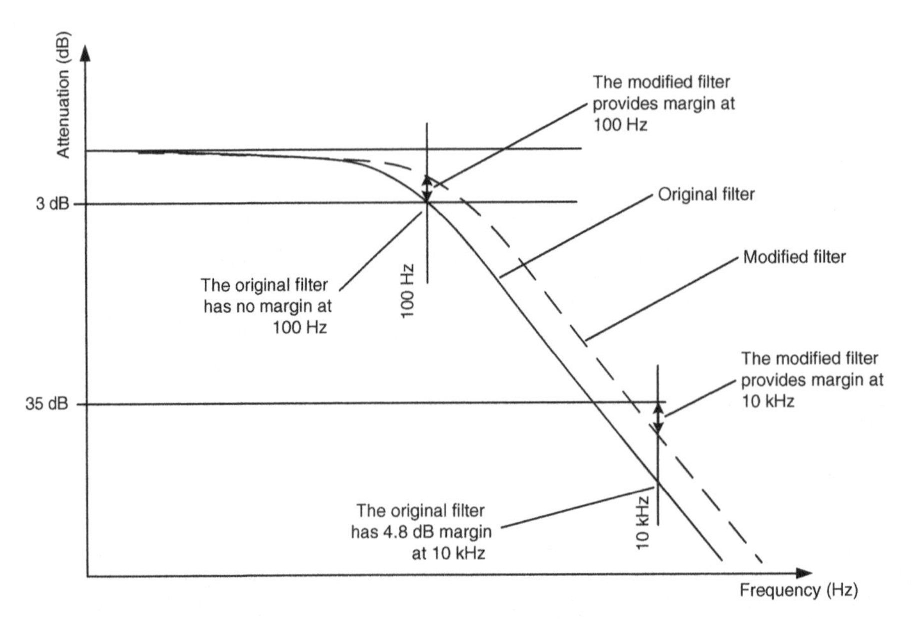

Figure 7.4. The original filter (solid line) has no margin at 100 Hz and excessive margin at 10 kHz. The modified filter (dashed line) exchanges margin at 10 kHz for margin at 100 Hz. Note this drawing is not to scale.

39.84 dB thereby *technically* satisfying the requirement. But noting that the 35 dB requirement is exceeded by 4.8 dB and the 3 dB requirement has no margin, we realize that if this filter were manufactured in quantity, component variations would result in an unacceptable failure rate.[5]

The 3 dB frequency is modified with the goal of reducing the margin at 10 kHz and increasing the margin at 100 Hz. Figure 7.4 shows that raising the filter's 3 dB frequency achieves this goal.

Substituting Equation 7.5 into Equation 7.3 provides a formula that allows us to optimize the 3 dB frequency:

$$\text{Atten}_{\text{dB}} = -\text{Gain}_{\text{dB}} = -20\log_{10}\left(\frac{1}{\sqrt{1+\left(\frac{f}{F_{3\text{dB}}}\right)^2}}\right) \tag{7.9}$$

Our goal is to have equal margins at 100 Hz and 10 kHz. We define the margin as Δ_{dB}. Our strategy is to write two equations in the two unknown

[5]Chapter 4 discusses manufacturing failure rates.

values Δ_{dB} and $F_{3\mathrm{dB}}$. Equations are written for the 100 Hz and 10 kHz cases.

$$3 - \Delta = -20\log_{10}\left(\frac{1}{\sqrt{1 + \left(\frac{100}{F_{3\mathrm{dB}}}\right)^2}}\right) \tag{7.10}$$

and

$$35 + \Delta = -20\log_{10}\left(\frac{1}{\sqrt{1 + \left(\frac{10{,}000}{F_{3\mathrm{dB}}}\right)^2}}\right) \tag{7.11}$$

Simultaneous solving of Equations 7.10 and 7.11 appears formidable but it can be done fairly easily. Our strategy is to add the two equations to eliminate Δ and then solve the resulting equation for $F_{3\mathrm{dB}}$:

$$38 = -20\log_{10}\left(\frac{1}{\sqrt{1 + \left(\frac{100}{F_{3\mathrm{dB}}}\right)^2}}\right) - 20\log_{10}\left(\frac{1}{\sqrt{1 + \left(\frac{10{,}000}{F_{3\mathrm{dB}}}\right)^2}}\right) \tag{7.12}$$

Rather than solve for $F_{3\mathrm{dB}}$ in Equation 7.12, we enter the equation in Excel and use the *Goal Seek*[6] function to solve for $F_{3\mathrm{dB}}$. The result is $F_{3\mathrm{dB}} = 151$ Hz and then Δ is calculated from Equation 7.10 or 7.11 as 1.42 dB. Therefore, the modified circuit has 1.42 dB margin for each requirement as shown in Figure 7.5.

7.3 FILTER RESPONSE CHARACTERISTICS

The lowpass RC filter discussed earlier has a *lowpass* response characteristic because it passes low frequencies and rejects high frequencies. Figure 7.6 shows the lowpass characteristic and the three other response characteristics: highpass, bandpass, and band-reject.

[6]If you don't have a symbolic equation solver, the Goal Seek function in Excel can save you hours of manipulation.

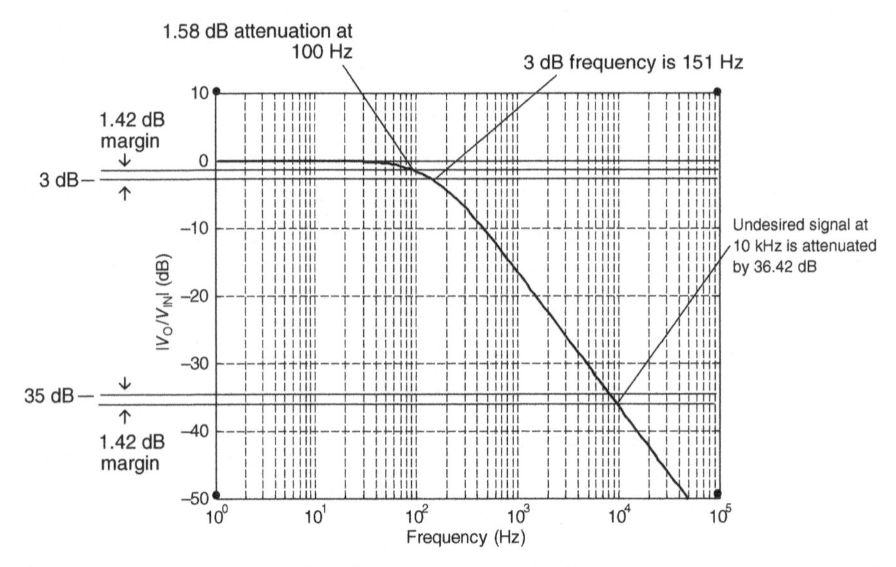

Figure 7.5. Magnitude response for Example 7.1. The filter has 1.42 dB margin at both 100 Hz and 10 kHz.

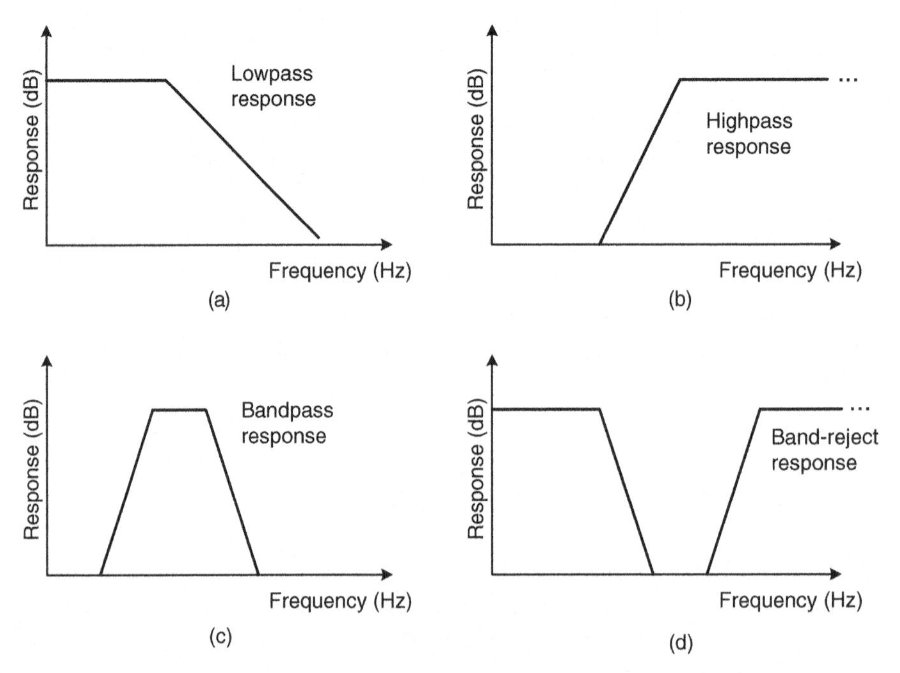

Figure 7.6. Filter response characteristics: lowpass (a), highpass (b), bandpass (c), band-reject (d).

7.4 SPECIFICATION OF FILTER TYPE

The lowpass RC filter is simple and easy to work with because it has only a single pole. By adding additional poles (and sometimes zeros), and manipulating their positions, different filter *types* can be created to satisfy specific performance requirements. Besides the lowpass RC, Butterworth, Chebyshev, and Bessel are common filter types. Some of their lowpass performance characteristics are described below and summarized in Table 7.1.

Passband flatness refers to the amount of amplitude variation at frequencies below the cutoff frequency. This is important when the filter must pass a signal consisting of multiple frequencies, and we don't want the filter to affect their relative magnitudes. Of the filter types above, the Butterworth filter provides maximum passband flatness.

Sharpness of response refers to the filter's ability to reject signals above the cutoff frequency. A very sharp filter will have 3 dB attenuation at the cutoff frequency and, say, 50 dB attenuation at twice the cutoff frequency. When a filter is used to reduce noise or remove an interfering signal, sharpness of response is often the dominant requirement. Of the filter types above, the Chebyshev filter has the sharpest response and the Bessel has the most gradual response.

Overshoot in step response refers to the filter's output when a step input is applied. A filter's step response can be critical when it is part of a system that must make measurements as quickly as possible. This is

TABLE 7.1. Comparison of performance characteristics for common filter types

	Main characteristic	Flatness in passband	Sharpness of response	Overshoot in step response	Flatness of group delay
Lowpass RC	Simple	Gradual roll-off	6 dB per octave roll-off	No overshoot	Peaks at cutoff frequency
Butterworth	Flat in passband	Flattest	Moderate	Moderate overshoot	Peaks near cutoff frequency
Chebyshev	Sharpest frequency response	Ripples in passband	Sharpest	Moderate overshoot	Peaks near cutoff frequency
Bessel	Flat group delay	Gradual roll-off	Moderate	No overshoot	Flattest passband group delay

common for devices that self-calibrate when first turned on. If the step response has overshoot, then more time must be allocated for the system to settle before making the measurement. Chebyshev and Butterworth filters both have overshoot in their step responses. The Bessel has no overshoot which makes it a good fit for measurement applications as will be shown in Example 7.3.

Flatness of group delay refers to how much the filter delays signals at different frequencies within its passband. If a signal consists of multiple frequencies, and if the different frequencies experience different amounts of delay, then the signal at the output would be distorted. This is important for broadband signals such as digital communication waveforms. The Bessel filter provides the flattest delay characteristic.

Being able to select the correct filter type could be your most valuable filter design skill. This section has presented some common filter types and requirements, but filter design textbooks show many more normalized plots that will give you even more flexibility with your design.[7] Understanding how to use normalized plots will help you "shop" for that perfect filter!

7.5 GENERALIZED FILTER DESIGN PROCEDURE

Filtering requirements vary widely. A filter design requirement may be lowpass, highpass, bandpass, or band-reject. If it is low pass or highpass, it will have a single cutoff frequency; if it is bandpass or band-reject, it will have two. It will also have a requirement for maximum attenuation in the passband, minimum attenuation in the stopband(s), and requirements specifying the steepness of the response between the passband and stopband(s). Fortunately modern filter design provides methods for converting this vast array of filtering requirements into a *normalized lowpass* requirement. With the filtering requirement converted into this form, a small number of precompiled plots and tables can be used to specify the filter type and order.[8] With the type and order specified, the filter can be designed from a filter table or by using an online tool. Our discussion is limited to lowpass filters, but the reader should be aware that designing highpass, bandpass, and band-reject filters is a straightforward extension of the material presented here and is well covered in [1].

[7]For example, a good compromise of group delay flatness *and* frequency roll-off is obtained using a filter that is Bessel to 6 dB and then follows a Chebyshev response. This is called a *transitional* filter [1].

[8]*Order* is related to the number of components used in the filter.

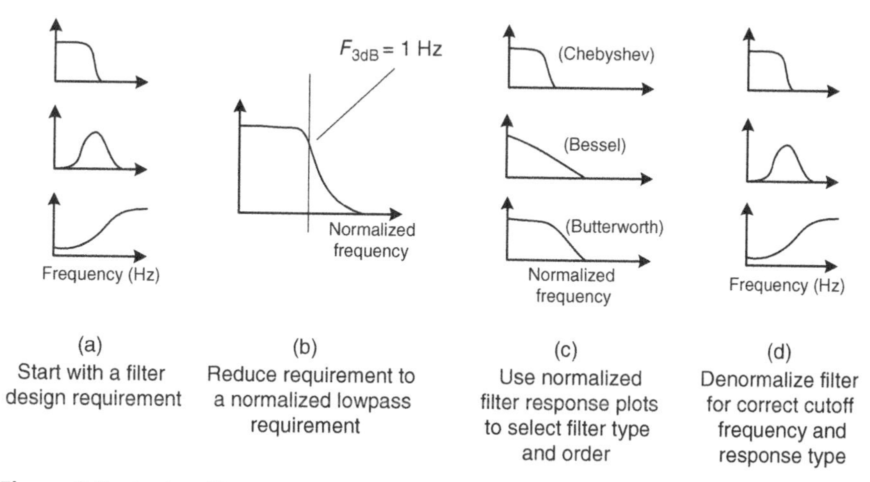

Figure 7.7. Analog filters are designed using the steps shown here and also outlined in the text. Our discussion is limited to lowpass filters, but references such as [1] show how to design filters with other response characteristics.

The steps for designing analog filters are described below and shown in Figure 7.7.

1. *Determine the desired response.* The desired response is usually a frequency-domain plot showing the desired attenuation as a function of frequency. It will have a lowpass, highpass, bandpass, or band-reject response characteristic.
2. *Convert the desired response to a normalized lowpass response.* For the lowpass filters used in this chapter, we normalize to the 3 dB frequency to 1 Hz. Though not covered here, filter design textbooks [1] show methods for converting highpass, bandpass, and band-reject requirements to normalized lowpass responses. The result of this step is the normalized lowpass requirement mentioned above and shown in Figure 7.7b.
3. *Use normalized response plots to determine the filter type and order.* The normalized lowpass frequency response for the filtering requirement is compared to normalized plots showing different filter types and orders. Based on this comparison, the required type and order are specified.
4. *Denormalize the filter and design using tables or an online resource.* With the filter type and order known, the filter is designed using the methods shown in the following sections.

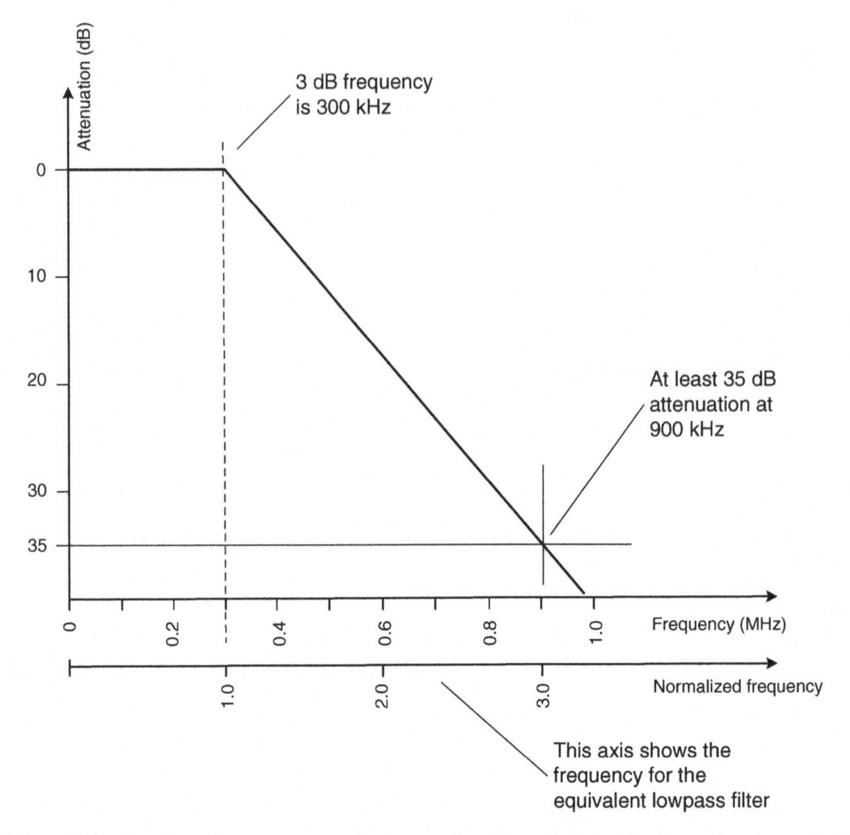

Figure 7.8. Frequency response requirements for Example 7.2. The lower horizontal axis shows frequency normalized to the desired 3 dB frequency.

Example 7.2. Select a filter type and order for a lowpass filter with 3 dB frequency of 300 kHz and 35 dB attenuation at 900 kHz. The response in the passband should be as flat as possible.

Solution. Referring to the steps above, the first step is complete because a lowpass filter is specified.

The second step is to scale the lowpass equivalent frequency response requirement so the 3 dB frequency is 1 Hz. This is done by dividing all frequency points by the 3 dB frequency as shown in Figure 7.8.

The normalized lowpass frequency response requirement is shown in Figure 7.9.

Since maximum flatness is desired, a Butterworth filter type is selected and the plot of Figure 7.10 is used to determine the required filter order. This plot shows that at a normalized frequency of 3, an $n = 4$ or fourth-order

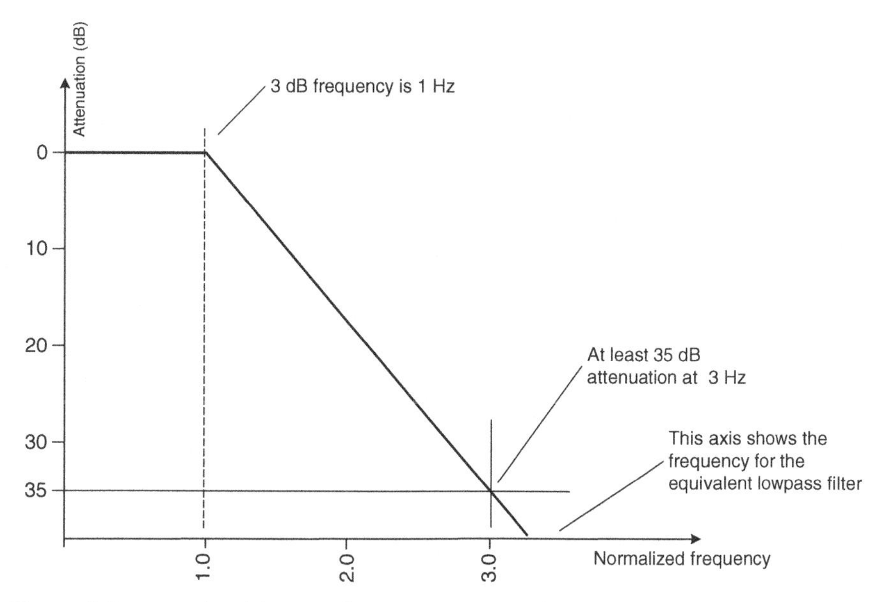

Figure 7.9. Normalized lowpass frequency response requirement for Example 7.2. This plot and Figure 7.10 are used to determine the required filter order.

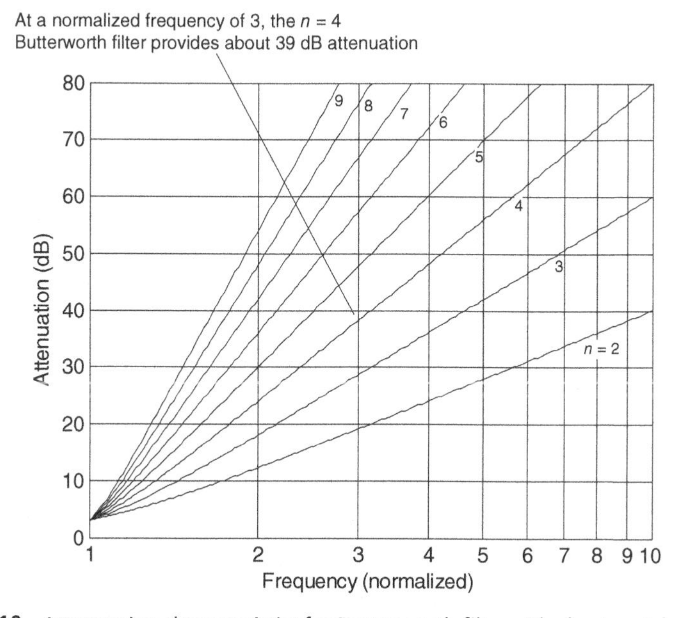

Figure 7.10. Attenuation characteristics for Butterworth filters. The horizontal axis is normalized frequency. The vertical axis is attenuation.

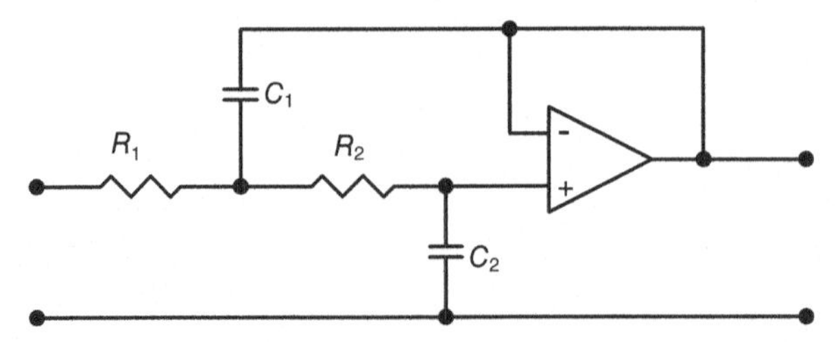

Figure 7.11. Sallen–Key lowpass active filter section. Similar but different topologies are used for highpass and bandpass filters.

Butterworth filter provides about 39 dB attenuation thereby satisfying the requirement. We conclude that the requirement is satisfied by a fourth-order Butterworth filter with cutoff frequency 300 kHz.

7.6 DESIGN OF ACTIVE LOWPASS FILTERS

There are several standard op-amp-based topologies that are used to implement active filters. The most popular is the Sallen–Key lowpass active filter section shown in Figure 7.11. Active filters are typically implemented by cascading one or more of these sections.

The transfer function for the filter shown in Figure 7.11 is

$$H(s) = \frac{1}{s^2 R_1 R_2 C_1 C_2 + sC_2 \left(R_1 + R_2\right) + 1} \tag{7.13}$$

Since the input impedance of the filter is typically high, and the output impedance is low, the transfer function of cascaded Sallen–Key structures is simply the product of the individual transfer functions. The transfer function is not needed for the filter design procedure, but is useful because it enables you to plot the frequency response of a filter once it has been designed.[9]

With the filter type and order specified, designing the filter is a straightforward process using equations found online or in textbooks. In Example 7.3 we use an online tool [2] to determine the components.

Example 7.3. A sensor, as shown in Figure 7.12, measures a process and produces a voltage of about 1 V, which is filtered and fed to an analog to digital converter (ADC) that samples the signal at 100 kHz. It is advantageous to sample the filter output as soon as possible after the signal appears at the

[9]This allows you to check your work as well and also evaluate component inaccuracies and so on.

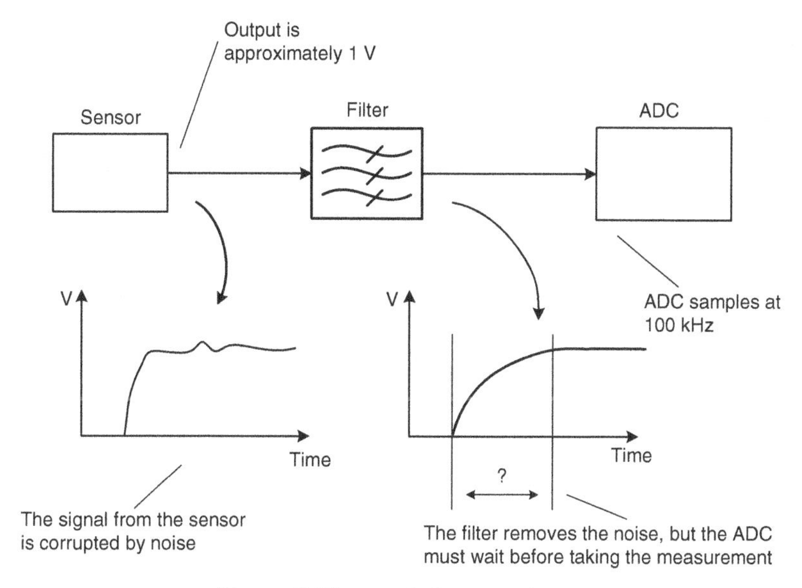

Figure 7.12. Details for Example 7.3.

sensor output. The filter is used to remove noise from the measurement signal and prevent noise above the Nyquist frequency of 50 kHz from aliasing into the passband when the signal is sampled. The 3 dB frequency of the filter must be 10 kHz and it must provide at least 40 dB attenuation at 50 kHz. Design the filter and specify the amount of time from when the signal is present at the sensor output to when it can be sampled by the ADC.

Solution. The first step is to check whether the lowpass RC filter will suffice. Equation 7.7 gives

$$\text{Octaves} = \frac{\ln\left(\frac{50\ \text{kHz}}{10\ \text{kHz}}\right)}{\ln(2)} = 2.32 \tag{7.14}$$

At 6 dB per octave the lowpass RC provides attenuation of only 13.9 dB at 50 kHz so it will not suffice.

Since measurement time is critical, we choose the Bessel filter type because its response to a step input has no overshoot. Since the 50–100 kHz frequency range for the filter is much lower than the bandwidth of most op-amps, and since the voltage of the signal to be measured is much higher than the millivolts of noise produced by most op-amps, an active filter will satisfy the requirement.

The requirement for a cutoff frequency of 10 kHz, and 40 dB attenuation at 50 kHz, means that the normalized response must have 40 dB attenuation at a

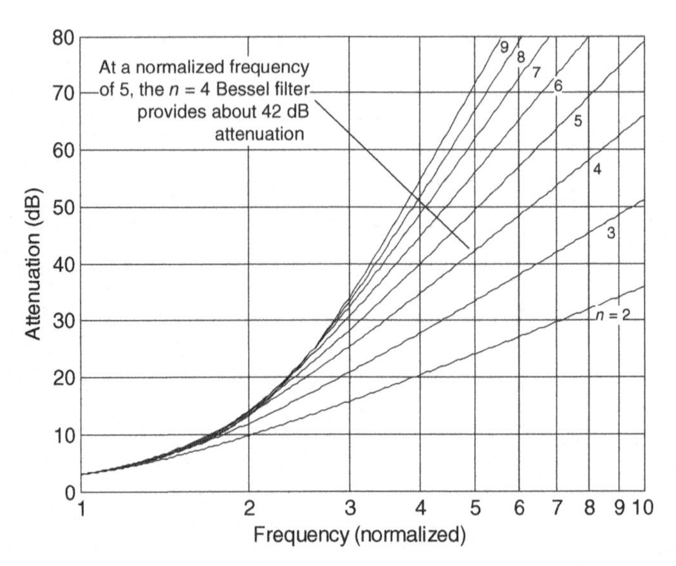

Figure 7.13. Attenuation characteristics for Bessel filters.

normalized frequency of 5. Figure 7.13 shows that a fourth-order Bessel filter will provide about 42 dB attenuation thereby exceeding the 40 dB requirement by 2 dB.

At this point the filter specification is complete. A fourth-order Bessel filter with 3 dB frequency of 10 kHz is required.

To specify the amount of delay from when the signal appears at the sensor output to when it can be sampled, we consult the step response plot in Figure 7.14. To use this plot, the time axis is denormalized by dividing by $2\pi f_c$, where f_c is the cutoff frequency. The denormalized axis is shown under the normalized axis. The curve for the fourth-order filter is essentially settled at a normalized time of 4 which denormalizes to 63.7 μs. Therefore, the system should wait 63.7 μs before making the measurement.

We have specified a fourth-order Bessel filter with cutoff frequency 10 kHz. The next step is to determine the components. The online tool in [2] is used to determine the components and the filter is shown in Figure 7.15.

We can verify the frequency response of the filter using the transfer function for individual sections from Equation 7.13. The transfer function of the two cascaded sections is shown in Equation 7.15, and the lines of Matlab in Figure 7.16 are used to create the plot in Figure 7.17:

$$H(s) = \frac{1}{s^2 R_1 R_2 C_1 C_2 + sC_2\left(R_1 + R_2\right) + 1} \cdot \frac{1}{s^2 R_3 R_4 C_3 C_4 + sC_4\left(R_3 + R_4\right) + 1} \tag{7.15}$$

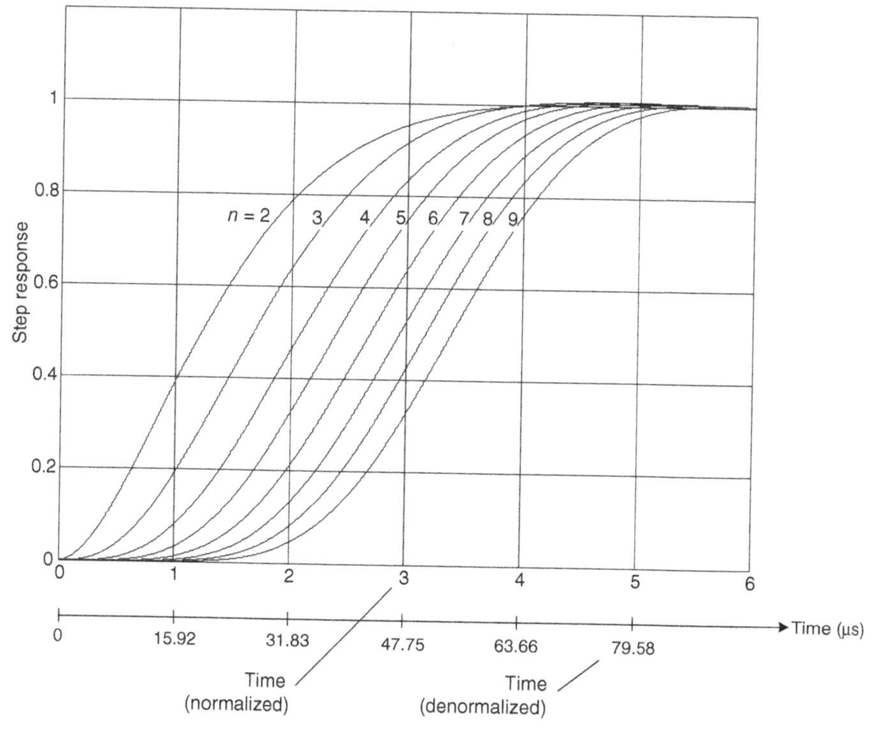

Figure 7.14. Step response plot for Bessel filters.

7.7 DESIGN OF PASSIVE RF FILTERS

Passive filters are typically more bulky and costly than active filters because they use inductors. However, in some applications passive filters are the only choice such as when a high-power RF amplifier generates harmonics that could interfere with other radio systems. These harmonics must be attenuated as shown in the following example.

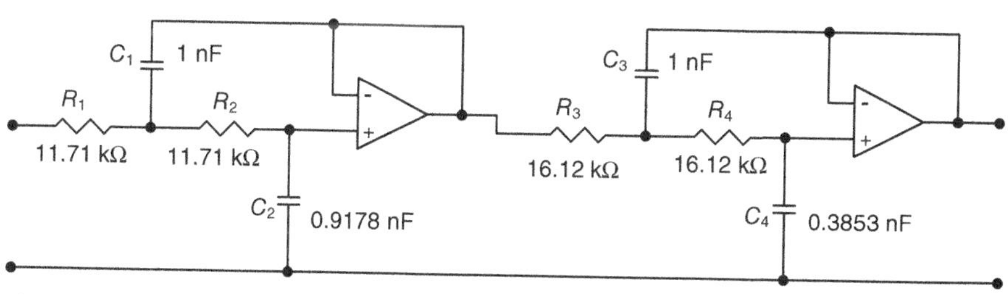

Figure 7.15. Fourth-order Bessel filter with 3 dB frequency 10 kHz. Component values were determined using the online tool [2].

```
1   StartFreq = 100;        % Lowest frequency to plot.
2   NumDec = 3;             % Number of decades to plot.
3   PtsPerDec = 200;        % Number of frequency points plotted per decade.
4
5   % Pre-allocate results matricies.
6   MagResp = zeros(NumDec*PtsPerDec,1);     % Matrix containing magnitude response.
7   Freq = zeros(NumDec*PtsPerDec,1);        % Matrix containing plot frequencies.
8
9   R1 = 11.71e3;           % Filter components
10  R2 = 11.71e3;
11  C1 = 1e-9;
12  C2 = 0.9178e-9;
13  R3 = 16.12e3;
14  R4 = 16.12e3;
15  C3 = 1e-9;
16  C4 = 0.3853e-9;
17
18  % Compute frequency response at frequency points uniformly-spaced on a log plot.
19  for i=1 : NumDec*PtsPerDec
20      Freq(i) = StartFreq*10^(i/PtsPerDec);    % Evaluate at this frequency.
21      s = 1j * 2*pi*Freq(i);                   % Determine complex frequency.
22      H1 = 1/(s^2*R1*R2*C1*C2 + s*C2*(R1+R2) + 1);    % Transfer function for stage
23      H2 = 1/(s^2*R3*R4*C3*C4 + s*C4*(R3+R4) + 1);    % Transfer function for stage
24      H = H1 * H2;                                     % Cascaded response
25      MagResp(i) = abs(H);        % Place magnitude response in result matrix.
26  end
27
28  % Plot frequency response in dB.
29  HDb = 20*log10(MagResp);
30  h11 = semilogx(Freq,HDb, 'color','k');
31  ylim([-50, 10]);
32  grid on;
33  xlabel('Frequency (Hz)');
34  ylabel('|Vo/Vin|, (dB)');
35  set(h11,'linewidth', 2);
```

Figure 7.16. Matlab code for plotting frequency response of active lowpass Bessel filter of Example 7.3. This method for plotting frequency response is described in Chapter 3.

Example 7.4. The power amplifier subsystem shown in Figure 7.18 is used to amplify signals between 7.5 and 10 MHz. The power amplifier outputs a squarewave[10] that is fed to a lowpass filter that attenuates harmonics of the fundamental, thereby producing a sinewave at the filter output. Design a filter that attenuates the fundamental less than 1 dB and all harmonics by at least 30 dB below the fundamental. The output impedance of the power amplifier is 50 Ω, and the filter is terminated in 50 Ω.

Solution. Prior to designing a filter to attenuate the harmonics, their magnitude must be known. If the peak amplitude of the waveform in Figure 7.18c is 1 V, then the waveform can be expanded as a Fourier series [3] as shown

[10]Power amplifiers sometimes output a square wave to increase their power efficiency thereby reducing power dissipation.

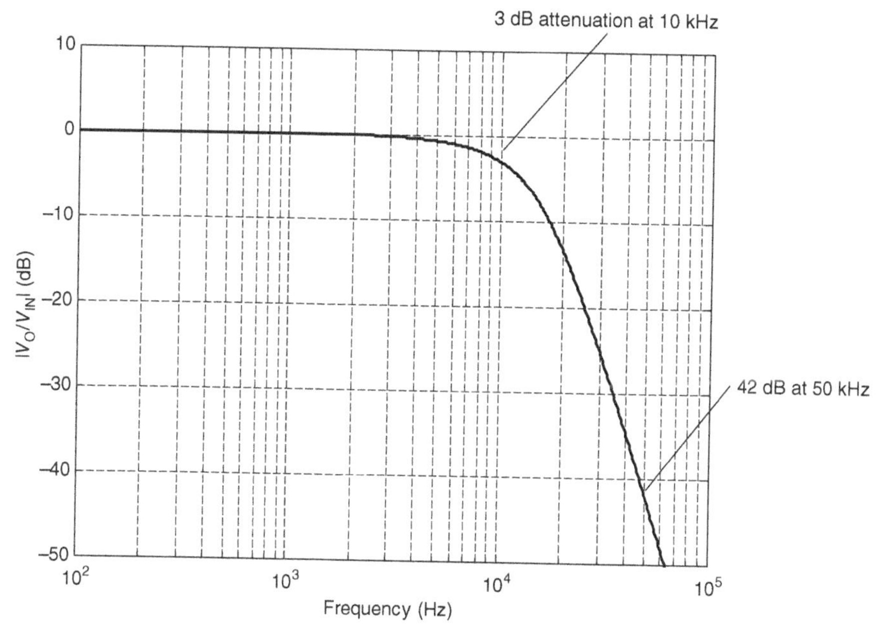

Figure 7.17. Computed frequency response for active lowpass Bessel filter of Example 7.3.

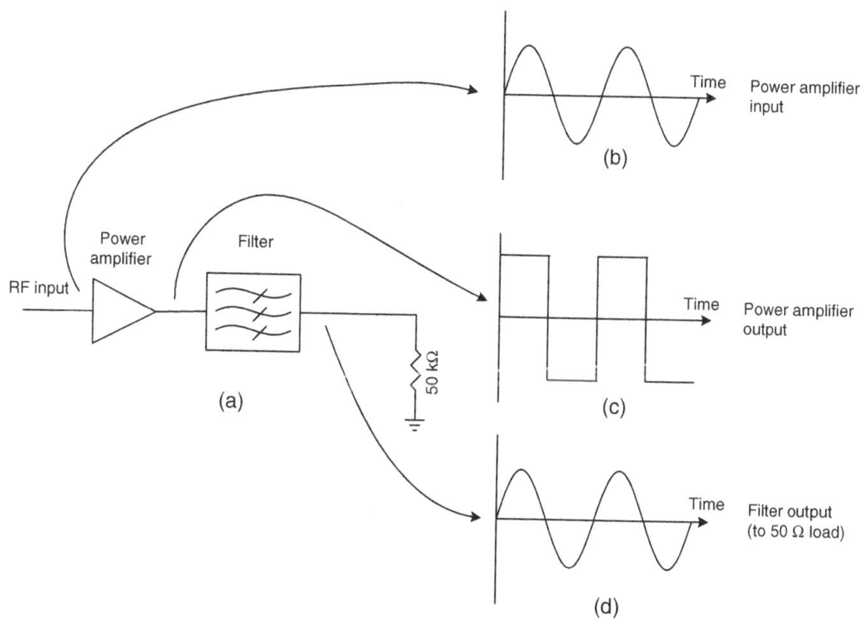

Figure 7.18. Power amplifier and filter for Example 7.4(a). The square-wave output from the amplifier (c) is filtered, leaving only the fundamental (d).

in Equation 7.16, where the fundamental is the first term of the series and all other terms represent unwanted harmonics:

$$f(\phi) = \frac{4}{\pi} \sum_{n=1,3,5,\ldots} \frac{1}{n} \sin(n\phi) \tag{7.16}$$

Equation 7.16 shows that the amplitude of the harmonics decreases with frequency. The third harmonic is closest in frequency to the fundamental so it drives the steepness requirement of the filter. From Equation 7.16 we see that the third harmonic is 1/3 the amplitude of the fundamental. Its attenuation in dB is

$$\begin{aligned}\text{Attenuation of third harmonic below fundamental} &= 20\log_{10}\left(\frac{1}{3}\right)\\ &= 9.54\text{dB}\end{aligned} \tag{7.17}$$

The filtering requirement is shown in Figure 7.19, where it is seen that the maximum attenuation is 1 dB at 10 MHz. If we assume the filter will

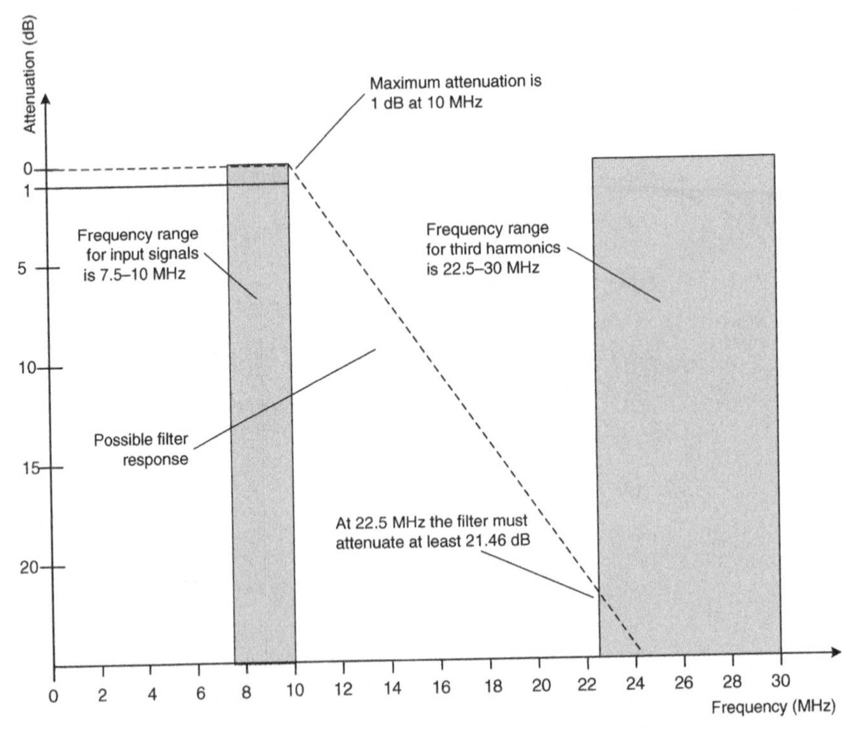

Figure 7.19. Filtering requirements for Example 7.4. The steepness requirement for the filter is driven by its response at 10 and 22.5 MHz.

attenuate a 7.5 MHz signal by no more than 1 dB, then for the third harmonic of 7.5 MHz to be 30 dB below the fundamental, the filter attenuation at 22.5 MHz must be

$$\text{Filter attenuation at 22.5 MHz} > 30\text{ dB} - 9.54\text{ dB} + 1\text{ dB} = 21.46\text{ dB} \tag{7.18}$$

Any of the filter types discussed earlier could satisfy this requirement. The Chebyshev will result in a lower order and consequently fewer parts, but it will have a small amount of ripple in the passband. The Bessel filter would provide flat group delay, but require more sections thereby using more parts. We use a Butterworth filter because it is a compromise between these choices. Attenuation characteristics for Butterworth filters are shown in Figure 7.20, which shows that for orders greater than 3, the attenuation of the Butterworth filter is less than 1 dB at a normalized frequency of about 0.8. We use this to specify the 3 dB frequency of the filter as follows:

$$f_{1\text{dB}} = 0.8 f_{3\text{ dB}} \tag{7.19}$$

so

$$f_{3\text{dB}} = \frac{f_{1\text{ dB}}}{0.8} = \frac{10\text{ MHz}}{0.8} = 12.5\text{ MHz} \tag{7.20}$$

and 22.5 MHz normalizes to

$$\text{Normalized frequency} = \frac{22.5\text{ MHz}}{12.5\text{ MHz}} = 1.8 \tag{7.21}$$

Therefore we must pick a filter order that has at least 21.46 dB attenuation at a normalized frequency of 1.8. Figure 7.20 shows that a fifth-order Butterworth filter would give about 25.7 dB thereby satisfying the requirement.

At this point the filter specification is complete. A fifth-order Butterworth filter with 3 dB frequency of 12.5 MHz is required. The components can be determined using normalized filter tables from [1] or from an online calculator [4]. Knowing that we will check the completed design by plotting the frequency response, we choose the latter approach[11] and design the filter shown in Figure 7.21 using [4].

[11]Online computational tools do not always produce correct results. This makes checking the result especially important.

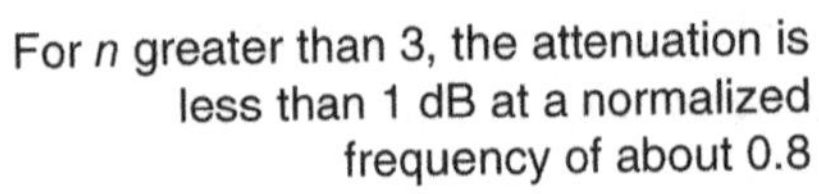

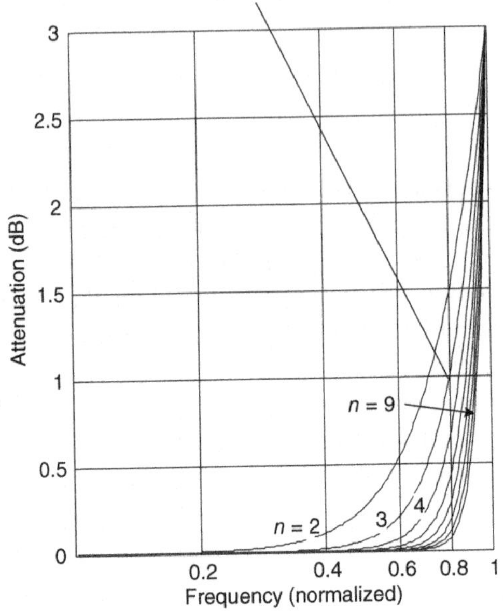

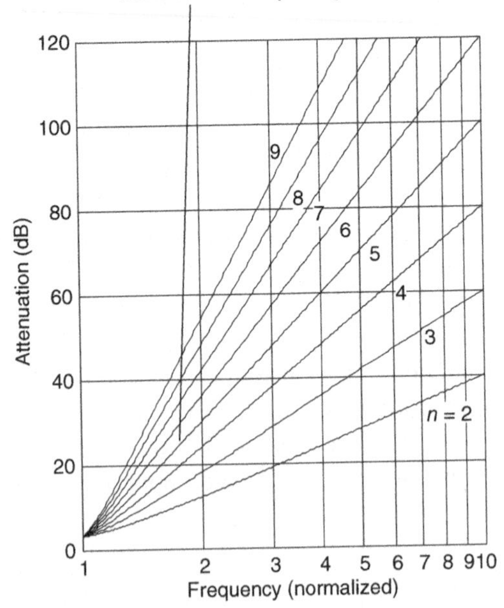

Figure 7.20. Attenuation characteristics for Butterworth filters.

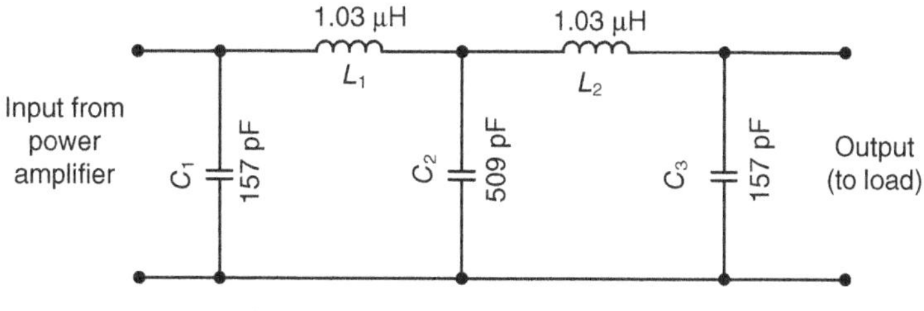

Figure 7.21. Filter for Example 7.4.

```
1    % See Williams, Page 2-3 and 2-4 for frequency and impedance scaling.
2    FSF = 12.5e6 * 2*pi;     % Frequency scaling factor
3    Z = 50;                  % Impedance scaling factor
4
5    % Normalized component values for 5th-order Butterworth lowpass filter.
6    % Williams (page 12-3)
7    C1N = 0.6180;
8    L2N = 1.6180;
9    C3N = 2;
10   L4N = 1.6180;
11   C5N = 0.6180;
12
13   % Compute actual values by frequency and impedance scaling the normalized
14   % values.
15   C1 = C1N/(FSF*Z);
16   L2 = L2N*Z/FSF;
17   C3 = C3N/(FSF*Z);
18   L4 = L4N*Z/FSF;
19   C5 = C5N/(FSF*Z);
```

Figure 7.22. Matlab code to compute component values from normalized values for Example 7.4.

The Matlab code in Figure 7.22 shows the process of determining the component value by frequency and impedance scaling the normalized filter values as shown in [1].

The filter's frequency response can be determined by analyzing the filter using *ladder analysis* as shown in Chapter 3. Figure 7.23 shows the currents

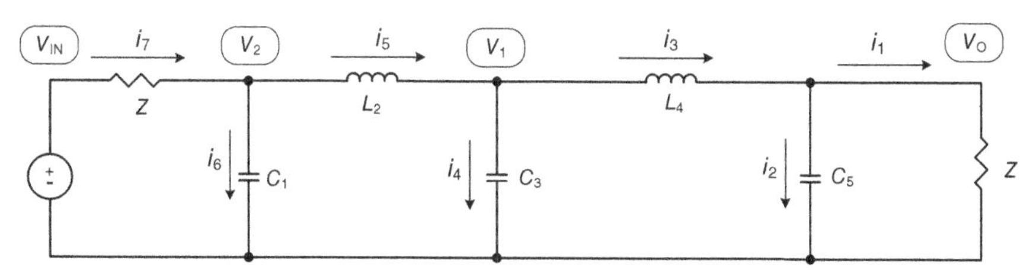

Figure 7.23. Analysis model for Example 7.4. Ladder networks are easy to analyze as shown in Figure 7.24.

```
1   NumDec = 3;                  % Number of decades to plot
2   PtsPerDec = 300;
3   StartFreq = 100000;     % Frequency in Hz
4   i = 1:NumDec*PtsPerDec;
5
6   F =  StartFreq*10.^(i/PtsPerDec);     % This is a logarithmic frequency sweep.
7   s = 1j*2*pi*F;                        % Define complex frequency.
8
9   % Compute impedances of each component at each frequency.
10  % Note the use of the dot operator which specifies element-wise operations
11  % in Matlab. For example,z1 is a vector containing the impedance of C1 at
12  % all frequencies.
13  z1 = 1./(s*C1);    z2 = s*L2;   z3 = 1./(s*C3);    z4 = s*L4;    z5 = 1./(s*C5);
14
15  % Analyze ladder network by setting output voltage to 1 volt and computing the
16  % input. The operations below create vectors of values.
17  Vo = 1;
18  i1 = Vo/50;
19  i2 = Vo./z5;
20  i3 = i1 + i2;
21  V1 = Vo + i3.*z4;
22  i4 = V1./z3;
23  i5 = i3 + i4;
24  V2 = V1 + i5.*z2;
25  i6 = V2./z1;
26  i7 = i5 + i6;
27  Vin = V2 + i7*Z;
28  H = Vo./Vin * 2;
29
30  % Plot frequency response.
31  HMag = abs(H);
32  HDb = 20*log10(HMag);
33  h11 = semilogx(F,HDb, 'color', 'k');
34  ylim([-40, 10]);
35  grid on;
36  xlabel('Frequency (Hz)');
37  ylabel('|Vo/Vin|, (dB)');
38  set(h11,'linewidth', 2);
```

Figure 7.24. Matlab code to determine frequency response for Example 7.4. The Matlab "dot operator" is used for element-wise vector operations in this code, so the "for" loop used in Figure 7.16 is not necessary.

and voltages used for the analysis done by the Matlab code in Figure 7.24. The frequency response is plotted in Figure 7.25.[12]

PROBLEMS

7.1 Determine the time constant and 3 dB frequency for the lowpass RC filter shown in Figure 7.26. If the capacitor is initially discharged and a step of 1 V is applied to the input, how long will it take for the output to reach 0.87 V?

[12]This analysis is useful for including the effects of inductor Q (which adds series resistance to the inductors) or the effects of component variation. See Problem 7.8.

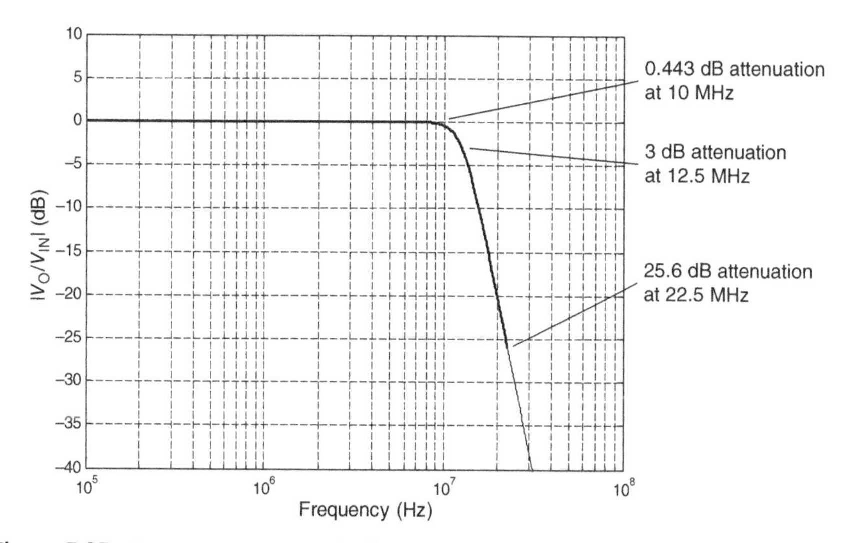

Figure 7.25. Frequency response plot for Example 7.4. This plot is produced by the Matlab code in Figure 7.24.

7.2 Determine the transfer function for the filter shown in Figure 7.27. Show that it is a highpass filter by substituting $s = j0$ and $s = j\infty$ into the transfer function. Show that the 3 dB frequency is $1/(2\pi RC)$.

7.3 Can an lowpass RC filter be used to provide 3 dB attenuation at 17 kHz and at least 30 dB attenuation at 625 kHz?

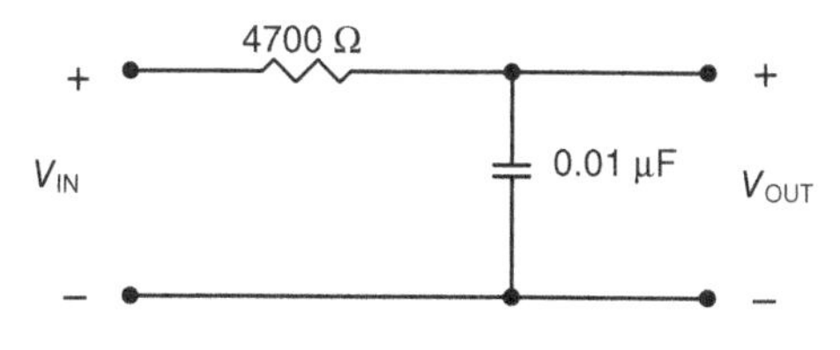

Figure 7.26. Circuit for Problem 7.1.

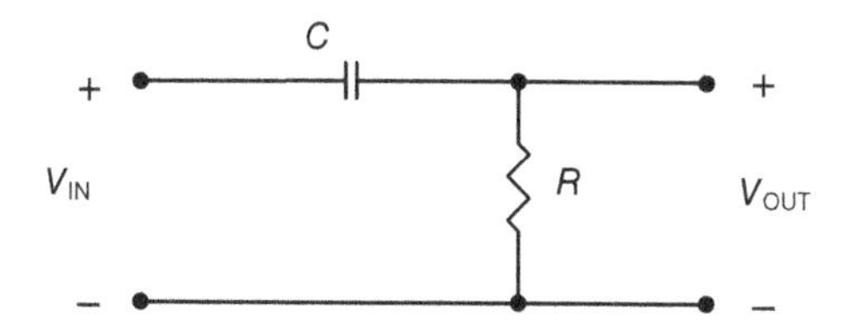

Figure 7.27. Highpass RC filter for Problem 7.2.

7.4 Can an lowpass RC filter be used to provide 1 dB attenuation at 21 kHz and at least 22 dB attenuation at 820 kHz?

7.5 A lowpass filter must have a 3 dB frequency of 13 MHz and provide at least 35 dB attenuation at 42 MHz. What is the normalized frequency corresponding to 42 MHz? Specify the order of a Butterworth filter that will satisfy this requirement.

7.6 A third-order Butterworth filter has 1 dB attenuation at 2800 Hz. What is its attenuation at 23.3 kHz?

7.7 Design a Sallen–Key lowpass Butterworth filter with a 3 dB bandwidth of 8 kHz and 20 dB attenuation at 24 kHz.

7.8 Design a fifth-order passive Chebyshev lowpass filter with cutoff frequency 21 MHz, impedance 50 Ω, and 0.1 dB passband ripple. Determine the effect on the filter's frequency response if the inductors have a quality factor of 10 at 21 MHz. *Hint:* The effect of inductor Q is a 7.78 Ω resistor in series with each inductor.

REFERENCES

1. A. B. Williams, *Electronic Filter Design Handbook*, McGraw-Hill, 1981.
2. Analog Devices Corporation. *Analog Filter Wizard*. Available: http://www.analog.com/en/amplifiers-linear/products/dt-adisim-design-sim-tool/filter_wizard/resources/fca.html (Accessed 12/1/2013.)
3. *CRC Standard Mathematical Tables*, 25th edn, CRC Press, 1979.
4. D. Heatherington. *LC Filter Designer*. Available: http://www.wa4dsy.net/filter/filterdesign.html (Accessed 12/1/2013.)

8

HOW TO DESIGN DIGITAL FILTERS

If your design requires a filter, and if the filter can be implemented in an existing field programmable gate array (FPGA) or microprocessor, then its effect on the product cost will be much less than if the filter were implemented using dedicated components such as capacitors, inductors, or op-amps. In addition, the design will be more reliable and smaller because it will have fewer components. Finally, it will not be affected by manufacturing tolerances, temperature, and aging. These advantages are compelling reasons to use digital filters wherever possible.

Digital filtering, or equivalently *discrete-time* filtering, is usually the best approach if the signal to be filtered can be represented as a set of uniformly spaced samples and a device is available to do the necessary computations. The continuing evolution of analog to digital converters (ADCs) and digital to analog converters (DACs) combined with advances in hardware- and firmware-based digital signal processors (DSPs) makes digital filtering more practical every day. As a working engineer, it is highly likely that you will find yourself designing digital filters or updating older designs to use them.

Ten Essential Skills for Electrical Engineers, First Edition. Barry L. Dorr.

Designing and debugging any kind of filter is fun and rewarding, but the process is very different for digital and analog filters. A digital filter is designed by selecting an architecture and then computing constants that determine the filter's bandwidth, cutoff frequency, and so on. Next, the filter's behavior is validated using a digital computer. When the filter is correctly implemented in either hardware or firmware, it will work *exactly* as you intended, and you'll spend minimal laboratory time troubleshooting. The scenario is different for analog filters. After the components have been selected, a printed circuit (PC) board must be designed and the components must be installed. When you test the prototype you may discover problems with component tolerances, PC layout effects, op-amp noise, inadequate component Q, and so on, and it may take some time to troubleshoot. If you have had experience troubleshooting a difficult analog circuit, you will certainly appreciate the accuracy and repeatability of digital filters![1]

This chapter includes quite a few equations, but you will quickly discover that the equations and z-transforms simplify the material rather than obscure it. As you read the chapter you will build a working knowledge of the z-transform and it will likely become a valuable part of your skill set. Section 8.1 describes how to convert analog signals to discrete-time signals and then how to convert discrete-time signals to analog signals. In Section 8.2 we show how to compute the z-domain transfer function of discrete-time structures and how to compute the frequency response of any z-domain transfer function. Next in Section 8.3 we review the commonly used finite impulse response (FIR) and infinite impulse response (IIR) filter structures and discuss where each would be an appropriate solution. The chapter concludes with Section 8.4 by reviewing several simple, but useful digital filters that are easy to analyze, design, and implement.

8.1 REVIEW OF SAMPLING

Digital filters operate on samples of continuous signals as shown in Figure 8.1. *Sampling* is the process of converting the continuous-time signal at the left side of Figure 8.1 to a discrete-time signal and then converting the digitally filtered signal back to a continuous-time signal. If the effects of sampling are not properly accounted for, your design may not work as expected.

Consider what happens if the input $s(t)$ to the system of Figure 8.1 is a sinusoid with frequency f:

$$s(t) = \cos(2\pi ft + \phi) \tag{8.1}$$

[1]You will see this at the end of the chapter where you design several practical and useful digital filters with nothing more than a standard spreadsheet program.

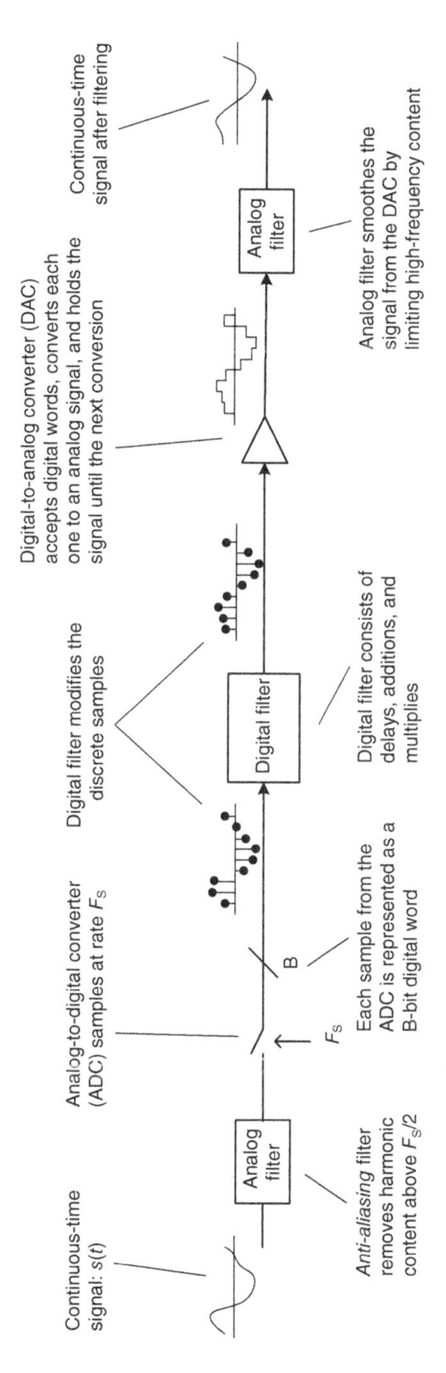

Figure 8.1. Steps for filtering a continuous-time signal with a digital filter.

$s(t)$ is sampled at $F_S = 1/T_S$ samples per second by the ADC. The resulting signal is

$$s(n) = \cos\left(2\pi f n T_S + \phi\right) = \cos\left(2\pi \frac{f}{F_S} n + \phi\right) \tag{8.2}$$

where n is the sample index.

To gain insight into the sampling process, we represent the frequency of the input signal, f, by

$$f = kF_S + \delta_f \tag{8.3}$$

where F_S is the sample rate, k is an integer, and δ_f is in the frequency in the range $-F_S/2$ to $F_S/2$.

Substituting Equation 8.3 into Equation 8.2 gives

$$s(n) = \cos\left(2\pi \frac{kF_S + \delta_f}{F_S} n + \phi\right) = \cos\left(2\pi kn + 2\pi \frac{\delta_f}{F_S} n + \phi\right) \tag{8.4}$$

The $2\pi kn$ term in Equation 8.4 simply adds a multiple of 2π to the phase angle of the term in parenthesis and can therefore be dropped. As a result

$$s(n) = \cos\left(2\pi \frac{\delta_f}{F_S} n + \phi\right) \tag{8.5}$$

Equation 8.5 shows the important result that regardless of the frequency of the input signal, the sampled signal will always be in the frequency range $-F_S/2$ to $F_S/2$. If input signals are within this range, then the frequency of the sampled signal will correspond to the frequency of the input signals. If not, they will be *aliased* into the range $-F_S/2$ to $F_S/2$ as shown graphically in Figure 8.2.[2]

In most cases, these aliased components are undesirable and are typically eliminated by an analog filter placed at the input of the sampler commonly called an *anti-aliasing* filter[3] and shown in Figure 8.1. In theory, the filter could be a *brick wall* filter as shown in Figure 8.3a. Since brick wall filters are unrealizable, signals are typically sampled at a rate greater than twice the

[2] In this chapter, you will see signals consisting of positive and negative frequencies. This is a consequence of using the Euler identities to describe sinusoids, for example, $\cos(\omega t) = \frac{e^{j\omega t} + e^{-j\omega t}}{2}$.

[3] Many commercially available ADCs include an anti-aliasing filter.

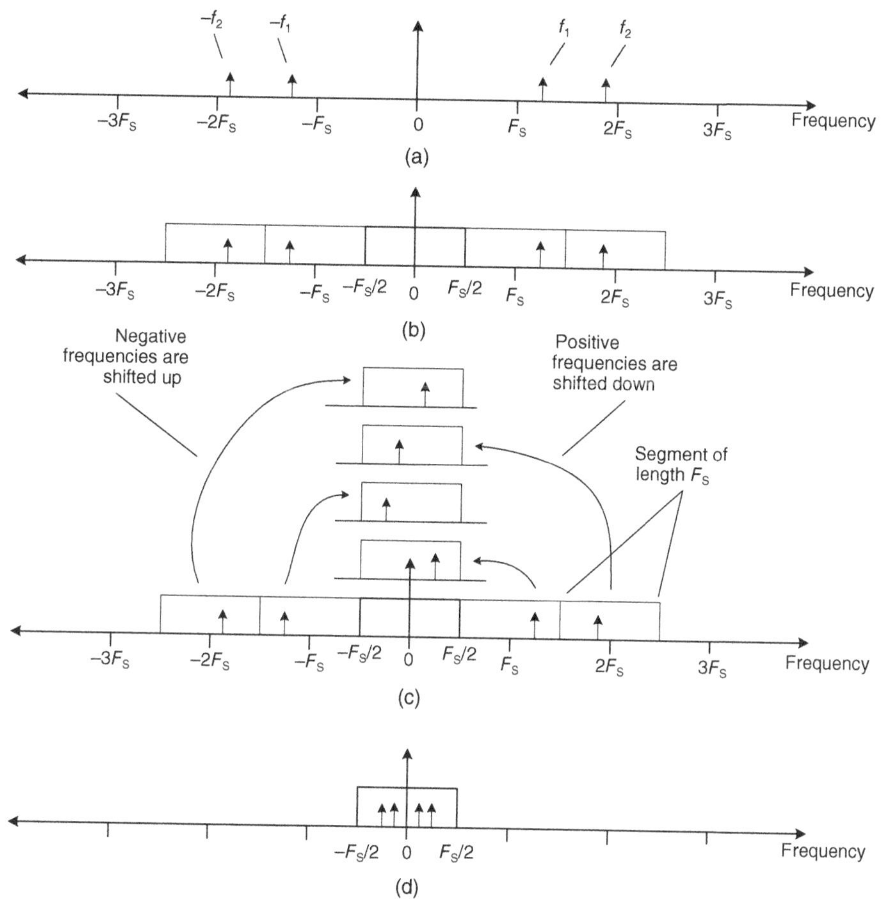

Figure 8.2. When signals outside the range $-F_S/2$ to $F_S/2$ are sampled, they are aliased into the range $-F_S/2$ to $F_S/2$. The continuous-time signal consists of two sinusoids at frequencies f_1 and f_2 (a). The frequency axis is divided into segments of length F_S (b). The sampling operation causes the original frequencies to be shifted by multiples of F_S to the range $-F_S/2$ to $F_S/2$ (c). The sampled signal consists of two frequency-shifted sinusoids (d).

Nyquist frequency or twice the highest frequency contained in the signal as shown in Figure 8.3b.[4]

The system of Figure 8.1 includes a DAC with effects that must be accounted for. The DAC accepts a discrete sample, converts it to an analog value, and holds the value constant until it receives the next sample.

[4]In this chapter we will sample signals at frequencies greater than twice the Nyquist frequency. However, aliasing can be used in useful ways by intentionally sampling signals below this rate as shown in [2].

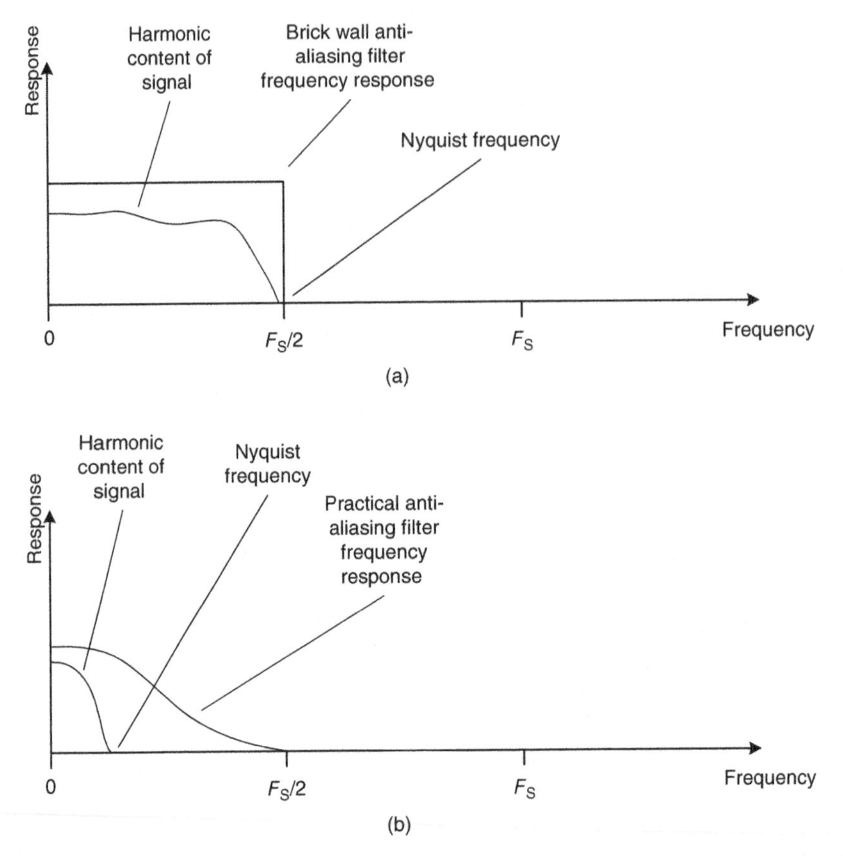

Figure 8.3. In theory aliasing can be eliminated by filtering the analog input signal with a "brick wall" filter that cuts off at the Nyquist frequency (a). Since brick wall filters are impractical, signals are often sampled at a frequency higher than twice the Nyquist frequency thereby allowing the anti-aliasing filter to have a more gradual roll-off characteristic (b).

Mathematically, the DAC, or *zero-order hold,* appears as a linear filter where the input is an impulse and the output is the value of the impulse held for one duration of the sampling clock. At first, this viewpoint seems strange, but elementary Laplace transforms can be used to show that the frequency response of this filter is the familiar sinc function

$$H(f) = \frac{\sin\left(\pi \frac{f}{F_S}\right)}{\left(\pi \frac{f}{F_S}\right)} \tag{8.6}$$

which has the effect of lowpass filtering the output signal. The process of converting a discrete-time signal to a continuous signal is shown in Figure 8.4.

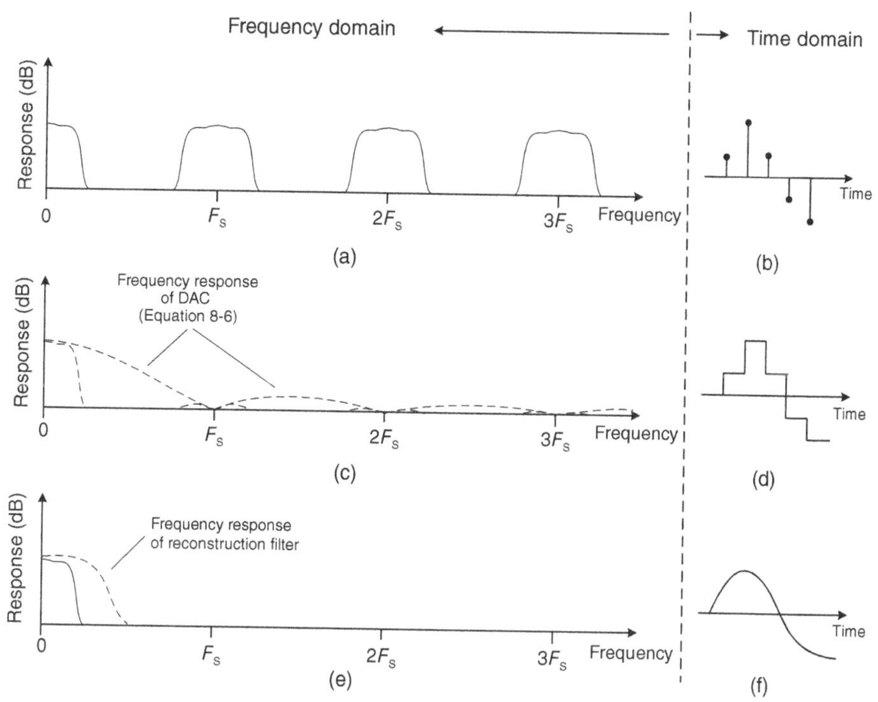

Figure 8.4. The signal at the DAC input is periodic in frequency and discrete in time (a, b). The DAC attenuates the signal harmonics by outputting a constant value during each sample time (c, d). The analog reconstruction filter further reduces the harmonics resulting in a continuous signal (e, f).

Since each sample causes an abrupt step at the DAC output, it contains harmonics as shown in Figure 8.4c. For this reason, an analog filter, called the *reconstruction filter*, placed at the DAC output, removes the excess harmonic content as shown in Figure 8.4e.

The effects of lowpass filtering by the DAC can be eliminated by digitally *pre-emphasizing* the digital signal with a response equal to the inverse of Equation 8.6. Some commercially available DACs provide pre-emphasis and/or a reconstruction filter.

8.2 USING THE Z-TRANSFORM TO DETERMINE THE TRANSFER FUNCTION AND FREQUENCY RESPONSE OF DIGITAL FILTERS

This section shows how to compute the z-domain transfer function for any discrete-time network and compute its frequency response. This will turn out

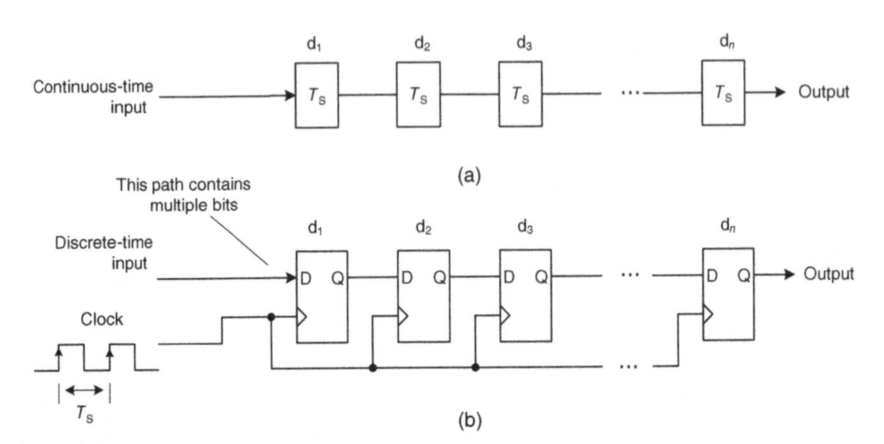

Figure 8.5. Continuous- and discrete-time cascades of delay elements. (a) The structure accepts a continuous-time input signal. (b) The structure accepts discrete samples of a signal. Both structures delay their input signal by nT_S seconds.

to be a valuable skill if you inherit an undocumented design or just want to check a design created from a filter design tool. This section will also reinforce the simplicity and utility of the z-transform.

Digital filters consist of three items: adders, multipliers, and delay elements—nothing else. Since adders and multipliers are not frequency selective, we correctly expect that the delay elements play a key role in the filter's frequency response. Figure 8.5a shows a cascade of continuous-time delay elements. Each element in Figure 8.5a delays its input by T_S seconds resulting in a total delay of nT_S seconds. Figure 8.5b shows a cascade of discrete-time delay elements. When a rising clock edge occurs, the signal at the input of each latch is transferred to its output. Signals at the left side of the structure are therefore shifted one position right every T_S seconds and the total delay is again nT_S seconds.

Intuitively we know that if the input of either structure in Figure 8.5 is a sinusoid, the output will be a phase-shifted replica of the input, and the phase shift from input to output will vary with frequency. Digital filters provide frequency selectivity by simply scaling and adding delayed signals or, equivalently, signals with frequency-dependent phases. We will now see that the z-transform allows us to account for the frequency-dependent phase shift in a convenient way.

The frequency response of an s-domain transfer function at frequency ω is obtained by substituting $j\omega$ for s and evaluating it. The transfer function of a delay of T_S seconds is given by the delay property of Laplace transforms

$$H(s) = \frac{Y(s)}{X(s)} = \mathrm{e}^{-sT_S} \tag{8.7}$$

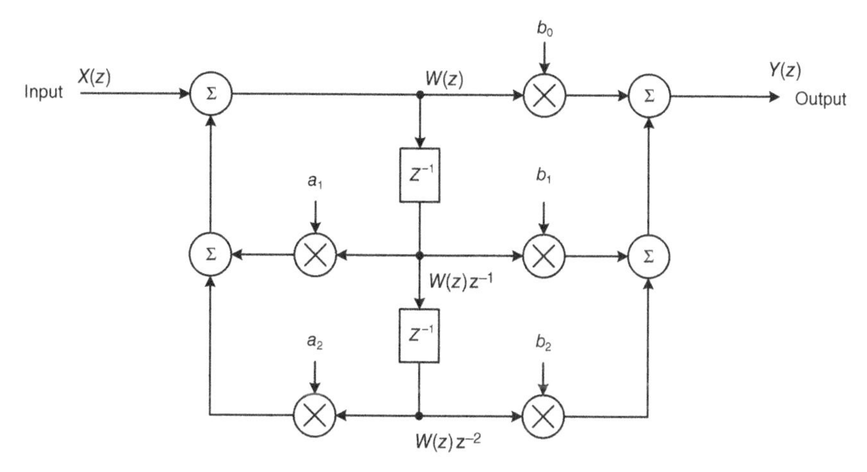

Figure 8.6. The direct form II filter structure provides a conjugate pair of poles and zeros and is a common building block for digital filters. The constants a_1, a_2, b_0, b_1, and b_2 establish the location of the poles and zeros and hence the frequency response of the filter.

which can be rewritten as

$$H(s) = \frac{Y(s)}{X(s)} = z^{-1} \tag{8.8}$$

where the frequency domain delay operator, z, is introduced as

$$z = e^{sT_S} = e^{j2\pi fT_S} \tag{8.9}$$

As mentioned before, digital filters consist only of adders, multipliers, and delay elements, and we have shown that the effect of a delay of n samples is represented in the frequency domain as a multiplication by z^{-n}. This allows us to express transfer functions of digital filters as functions of z as shown in Example 8.1.[5]

Example 8.1. Determine the z-domain transfer function of the *direct form II*[6] *filter* section shown in Figure 8.6.

Solution. We begin by noting that the signal at the input of the upper latch is $W(z)$. Therefore the output of the upper latch is $W(z)z^{-1}$ and the output of

[5] A more intuitive description of the z operator is provided in Chapter 5.

[6] This structure is commonly used for digital implementations of analog filter architectures.

the lower latch is $W(z)z^{-2}$. This allows us to express $W(z)$ as

$$W(z) = X(z) + W(z)z^{-1}a_1 + W(z)z^{-2}a_2 \tag{8.10}$$

resulting in the intermediate transfer function

$$\frac{W(z)}{X(z)} = \frac{1}{1 - a_1 z^{-1} - a_2 z^{-2}} \tag{8.11}$$

We now write the expression for the output, $Y(z)$, as a function of $W(z)$:

$$Y(z) = W(z)b_0 + W(z)z^{-1}b_1 + W(z)z^{-2}b_2 \tag{8.12}$$

Or

$$\frac{Y(z)}{W(z)} = b_0 + z^{-1}b_1 + z^{-2}b_2 \tag{8.13}$$

Equations 8.11 and 8.13 can be combined to give the z-domain transfer function of the structure:

$$H(z) = \frac{Y(z)}{X(z)} = \frac{Y(z)}{W(z)} \cdot \frac{W(z)}{X(z)} = \frac{b_0 + b_1 z^{-1} + b_2 z^{-2}}{1 - a_1 z^{-1} - a_2 z^{-2}} \tag{8.14}$$

The previous example showed how the z-transform enables us to write transfer functions for digital filters. To determine the frequency response of a z-domain transfer function, $H(z)$, we follow the steps shown in Figure 8.7.

Example 8.2. The direct form II filter section shown in Figure 8.8 has a sample rate of 240 kHz and coefficients as shown. Compute the frequency response.

Solution. Substituting the coefficients from Figure 8.8 into the z-domain transfer function of the direct form II filter section from Equation 8.14 gives

$$H(z) = \frac{0.0147 - 0.0071z^{-1} + 0.0147z^{-2}}{1 - 1.8177z^{-1} + 0.8428z^{-2}} \tag{8.15}$$

The lines of Matlab code shown in Figure 8.9 follow the process outlined in Figure 8.7. The magnitude and phase response are shown in Figure 8.10.

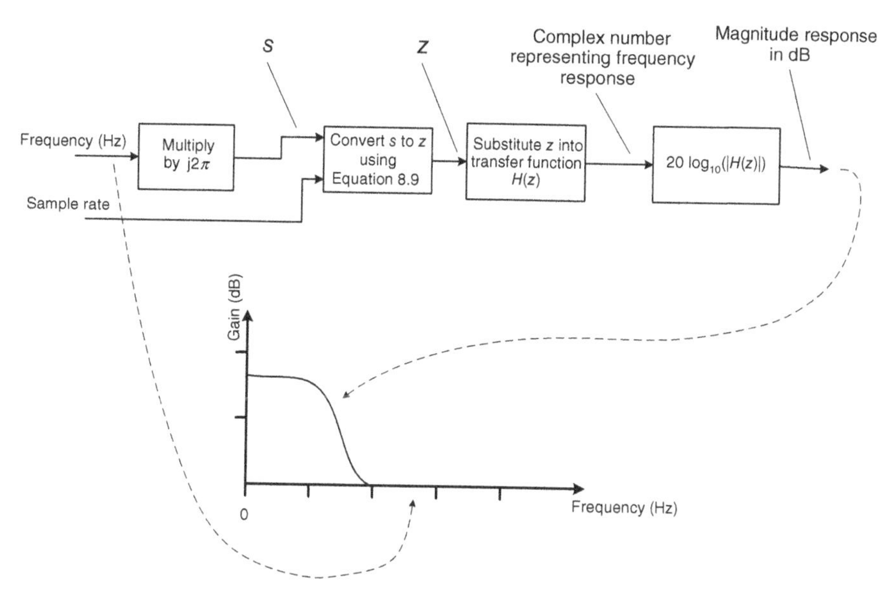

Figure 8.7. Illustration showing the steps for computing the frequency response of any z-domain transfer function.

There is an important difference between evaluating the frequency response of an analog filter and a digital filter. For the analog filter, every value of s produces a unique value of frequency response; for example, as the frequency increases, the attenuation of a lowpass RC filter will fall by 6 dB per octave indefinitely. But as the frequency increases in Equation 8.9, we see

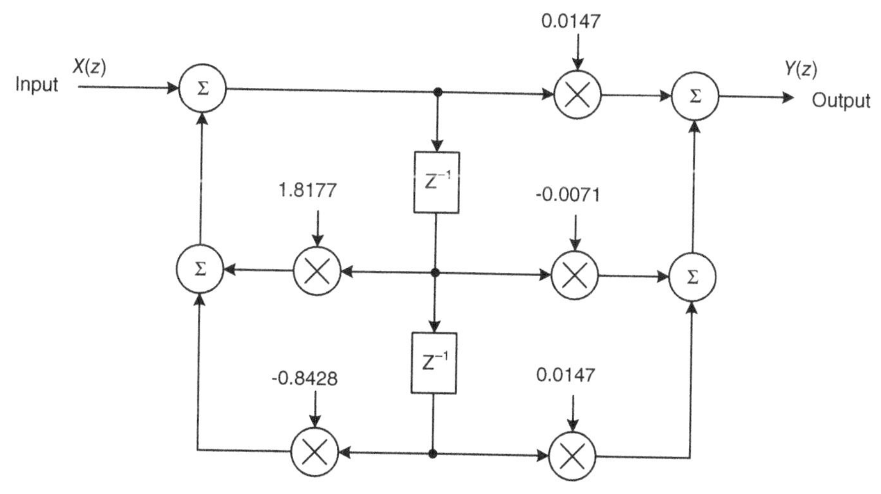

Figure 8.8. Direct form II filter section for Example 8.2.

```
1    a = [1 1.8177 -0.8428];              % Constants for denominator of transfer function.
2    b = [0.0147 -0.0071 0.0147];         % Constants for numerator of transfer function.
3
4    Fs = 240000;              % Sampling frequency in Hz.
5    StartFreq = 1000;         % Lowest frequency to plot.
6    StopFreq = Fs;            % Highest frequency to plot.
7    NumPts = 10000;           % Number of points to plot.
8    FreqSp = (StopFreq - StartFreq)/(NumPts-1); % Frequency spacing.
9
10   % Pre-allocate results matricies.
11   MagResp = zeros(NumPts,1);       % Matrix containing magnitude response.
12   PhaseResp = zeros(NumPts,1);     % Matrix containing phase response.
13   Freq = zeros(NumPts,1);          % Matrix containing plot frequencies.
14
15   % Compute frequency response at uniformly-spaced frequency points.
16   for i=1 : NumPts
17       Freq(i) = StartFreq + (i-1)*FreqSp;        % Evaluate at this frequency.
18       z = exp(1j*2*pi*Freq(i)/Fs);               % Define z using Equation 9.
19
20       % Equation 14 - z-domain transfer function for Direct Form II section.
21       H = (b(1) + b(2)*z^-1 + b(3)*z^-2)/(1 - a(2)*z^-1 - a(3)*z^-2);
22
23       MagResp(i) = abs(H);                       % Place magnitude response in result matrix.
24       PhaseResp(i) = angle(H)*360/(2*pi);        % Place phase response in result matrix.
25   end
26
27   % Plot magnitude and phase.
28   subplot(2,1,1);
29   HMag=semilogx(Freq,20*log10(MagResp), 'color','k', 'linewidth', 2); % Plot magnitude
30   set(gca,'FontSize',12);
31   xlabel('Freq (Hz)');
32   ylabel('Response (dB)');
33   xlim([0 Fs/2]);
34   ylim([-50 10]);
35   grid on;
36
37   subplot(2,1,2);
38   HPhs = semilogx(Freq,PhaseResp, 'color','k', 'linewidth', 2); % Plot phase
39   set(gca,'FontSize',12);
40   xlabel('Freq (Hz)');
41   ylabel('Phase (Degrees)');
42   xlim([0 Fs/2]);
43   ylim([-180 180]);
44   set(gca,'YTick',[-180, -90, 0, 90, 180]);
45   grid on;
```

Figure 8.9. Matlab code used to plot frequency response for *z*-domain transfer functions.

that the resulting value of z follows a circular trajectory for a digital filter because

$$z = e^{jsT_S} = \cos\left(2\pi f T_S\right) + j\sin\left(2\pi f T_S\right) \tag{8.16}$$

As a result, evaluating the frequency response of the z-domain transfer function at $f_2 = f_1 + kF_S$, where k is any integer, will give the same result as evaluating it at f_1, and we conclude that the frequency response of a z-domain transfer function is periodic with period $F_S = 1/T_S$. The relationship between the s-plane and z-plane is shown in Figure 8.11.

We are often interested in the DC response of digital filters, especially if the filter is lowpass. The DC response can be easily computed by noting that when $f = 0$, $s = 0$, and $z = e^{sT_S} = 1$. Therefore we can compute the DC response of any z-domain transfer function by substituting $z = 1$.

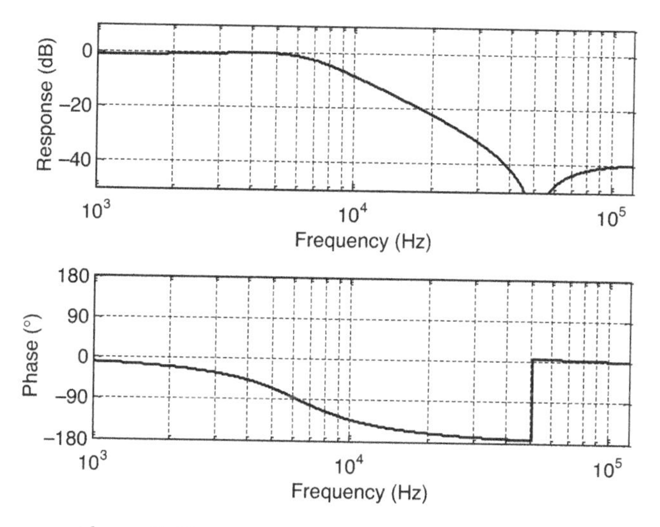

Figure 8.10. Frequency response for Example 8.2.

8.3 FIR AND IIR DIGITAL FILTERS

Most of the digital filters we encounter will be either FIR or IIR. The impulse response of a digital filter is the filter output resulting from setting the output of all internal latches to zero, applying a single nonzero input sample and then applying additional zeros as the filter is repeatedly clocked.

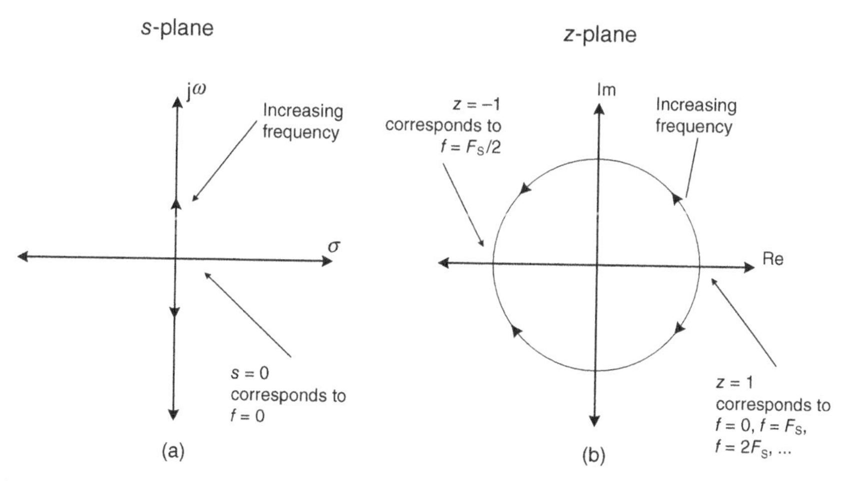

Figure 8.11. Frequency response of an *s*-domain transfer function is evaluated by varying *s* along the *jω* axis (a). Frequency response of a *z*-domain transfer function is evaluated by varying *z* around the unit circle (b) and is therefore periodic with period equal to the sample rate.

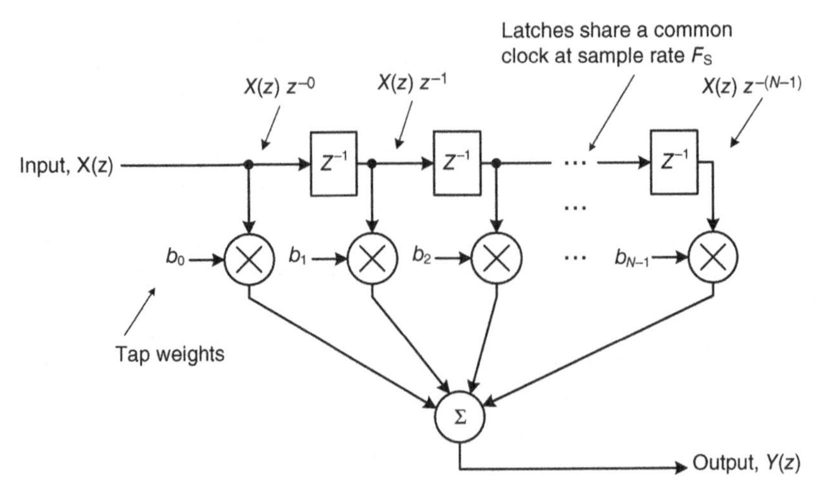

Figure 8.12. FIR filter structure. The latches shown by the blocks marked z^{-1} are clocked at the sample rate F_S.

8.3.1 FIR Filters

The FIR filter takes its name from the fact that its impulse response is limited to the length of the filter register. This can be seen in Figure 8.12 where an input signal enters at the left side of the filter and is shifted one position right each time the filter is clocked. Once the input reaches the right side of the filter it has no further effect. Since there is no feedback from output to input, FIR filters are unconditionally stable.

The input signal in Figure 8.12 is represented by $X(z)$. The output of the first latch is its input multiplied by z^{-1}, the output of the second latch is $X(z)z^{-2}$, and so on. The z-domain representation of the filter output is therefore

$$Y(z) = X(z)z^0 b_0 + X(z)z^{-1} b_1 + \cdots + X(z)z^{-(N-1)} b_{(N-1)} \tag{8.17}$$

which can be rewritten as

$$Y(z) = X(z) \sum_{n=0}^{N-1} b_n z^{-n} \tag{8.18}$$

And the z-domain transfer function of the FIR filter is

$$H(z) = \frac{Y(z)}{X(z)} = \sum_{n=0}^{N-1} b_n z^{-n} \tag{8.19}$$

FIR filters are often used because they can be designed so the phase shift caused by the filter is a linear function of frequency or equivalently, the filter delays all frequencies equally.[7] Linear phase, or flat *group delay*, is highly desirable for signals composed of multiple frequencies, such as those used for communication waveforms, because it minimizes distortion. An FIR filter has linear phase if the filter tap weights are symmetric, meaning that the taps on the right-hand side of the filter are a mirror image of the taps on the left.

When the taps are symmetric, as is often the case, the frequency response can be computed without having to manipulate complex numbers. This enables you to plot frequency response with a standard spreadsheet program. Consider an FIR with an odd number of symmetric taps. Multiplying the transfer function of Equation 8.19 by $z^{-N/2}e^{sT_S}$[8] and indexing the $N+1$ tap weights from $-N/2$ to $N/2$ allows us to write Equation 8.19 as

$$H(z) = \sum_{n=-N/2}^{N/2} b_n z^{-n} \tag{8.20}$$

Since the taps are symmetrical $b_k = b_{-k}$ and Equation 8.20 can be written as

$$H(z) = b_0 z^0 + \sum_{n=1}^{N/2} b_n (z^n + z^{-n}) \tag{8.21}$$

The frequency response is computed by substituting e^{sT_S} for z in Equation 8.21:

$$H(s) = b_0 + \sum_{n=1}^{N/2} b_n \left(e^{snT_S} + e^{-snT_S}\right) \tag{8.22}$$

Using the Euler identity $e^{j\phi} + e^{-j\phi} = 2\cos(\phi)$ and substituting $s = 2\pi f$ gives the equation for the frequency response:

$$H(f) = b_0 + 2\sum_{n=1}^{N/2} b_n \cos\left(2\pi \frac{f}{F_S} n\right) \tag{8.23}$$

[7]If you don't understand this sentence, consider a sinewave that encounters a delay, $\sin(\omega(t-\tau)) = \sin(\omega t - \omega\tau) = \sin(\omega t - \phi)$ where $\phi = \omega\tau$. Hence the phase shift is a linear function of frequency.

[8]Multiplying or dividing a z-domain transfer function by z will affect its phase response but not its amplitude response because the magnitude of the factor $e^{(sT_S)}$ is unity at all frequencies.

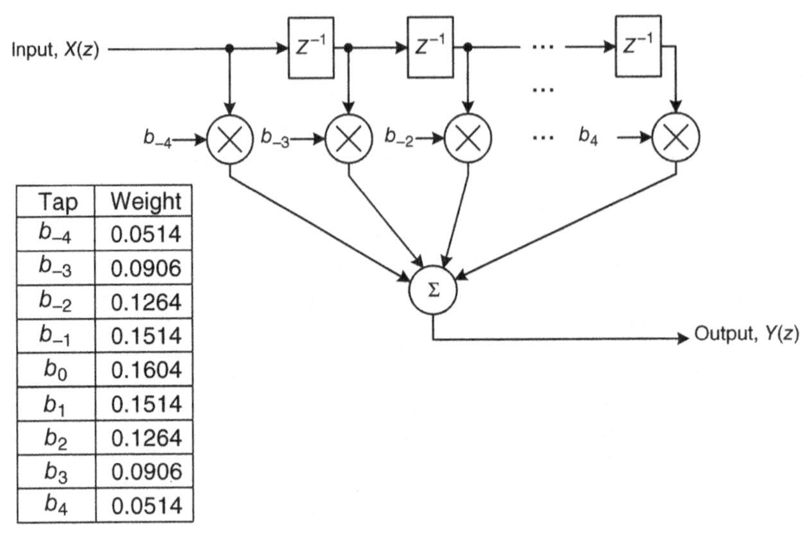

Tap	Weight
b_{-4}	0.0514
b_{-3}	0.0906
b_{-2}	0.1264
b_{-1}	0.1514
b_0	0.1604
b_1	0.1514
b_2	0.1264
b_3	0.0906
b_4	0.0514

Figure 8.13. FIR filter for Example 8.3.

Example 8.3. Plot the frequency response of the FIR filter shown in Figure 8.13. The sample rate is 10 kHz.

Solution. The table in Figure 8.13 shows that there is an odd number of taps that are symmetric about the center tap. Therefore the frequency response can be computed using Equation 8.23. We use Excel to compute the response as shown in Figure 8.14. Figure 8.15 shows the frequency response.

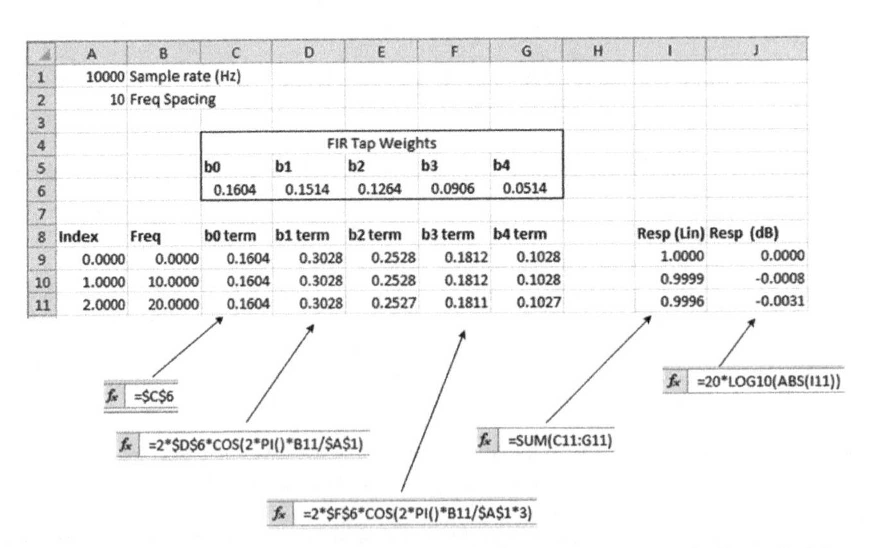

	A	B	C	D	E	F	G	H	I	J
1	10000	Sample rate (Hz)								
2	10	Freq Spacing								
3										
4			FIR Tap Weights							
5			b0	b1	b2	b3	b4			
6			0.1604	0.1514	0.1264	0.0906	0.0514			
7										
8	Index	Freq	b0 term	b1 term	b2 term	b3 term	b4 term		Resp (Lin)	Resp (dB)
9	0.0000	0.0000	0.1604	0.3028	0.2528	0.1812	0.1028		1.0000	0.0000
10	1.0000	10.0000	0.1604	0.3028	0.2528	0.1812	0.1028		0.9999	-0.0008
11	2.0000	20.0000	0.1604	0.3028	0.2527	0.1811	0.1027		0.9996	-0.0031

Figure 8.14. Excel worksheet used to compute frequency response in Example 8.3.

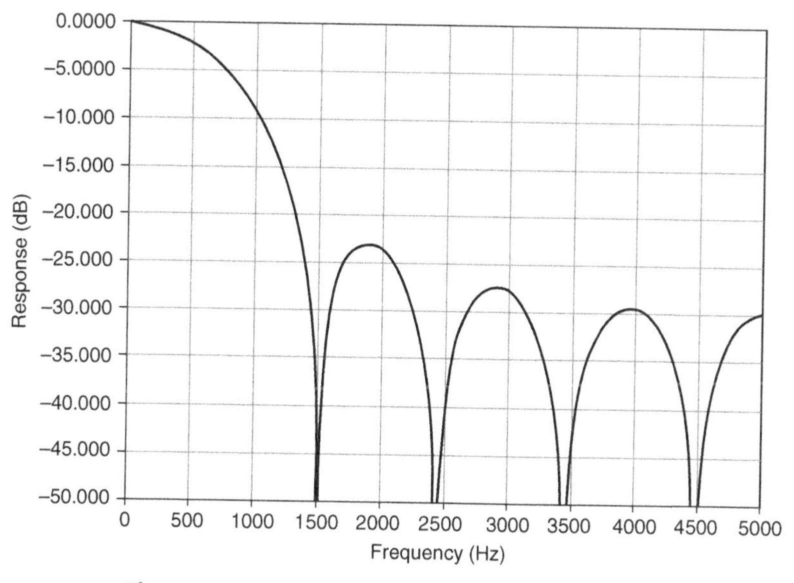

Figure 8.15. Frequency response plot for Example 8.3.

A formula similar to Equation 8.23 can be used when the number of taps, N, is even. In this case, the $N/2$ taps on the right side of the filter are a mirror image of the $N/2$ taps on the left. If the taps on the right are indexed from 1 to $N/2$, then the formula for frequency response is

$$H(f) = 2\sum_{n=1}^{N/2} b_n \cos\left(2\pi \frac{f}{F_S}(n - 0.5)\right) \tag{8.24}$$

8.3.2 IIR Filters

The IIR filter takes its name from the fact that its impulse response continues indefinitely. This is seen in the simple IIR of Figure 8.16, where the latch is initialized to zero and then a single sample of unity is fed to the input followed by all zeros. Since the output is multiplied by the feedback coefficient 0.5, successive clock cycles will produce output values of 0.5, 0.25, 0.125, and so on. The output will become vanishingly small, but theoretically will not entirely disappear.[9] The reason that the impulse response is infinite is that the filter has feedback from output to input. Therefore IIR filters have the potential to become unstable if not designed correctly.

[9]In a practical filter, the output will eventually decrease below the bit-precision of the filter and disappear entirely.

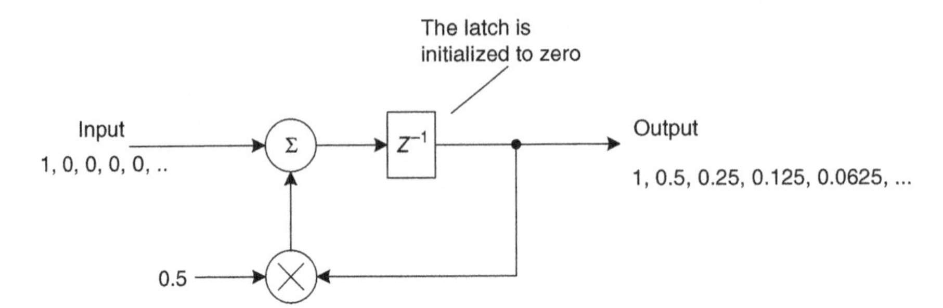

Figure 8.16. If the latch is initially preloaded with zero and a single one is applied, the output of the IIR filter continues indefinitely.

Figure 8.17 shows the general structure of an IIR filter with input $X(z)$ and output $Y(z)$. To compute the z-domain transfer function we first compute the output

$$Y(z) = X(z) + Y(z)z^{-1}a_1 + Y(z)z^{-2}a_2 + \cdots + Y(z)z^{-N}a_N \qquad (8.25)$$

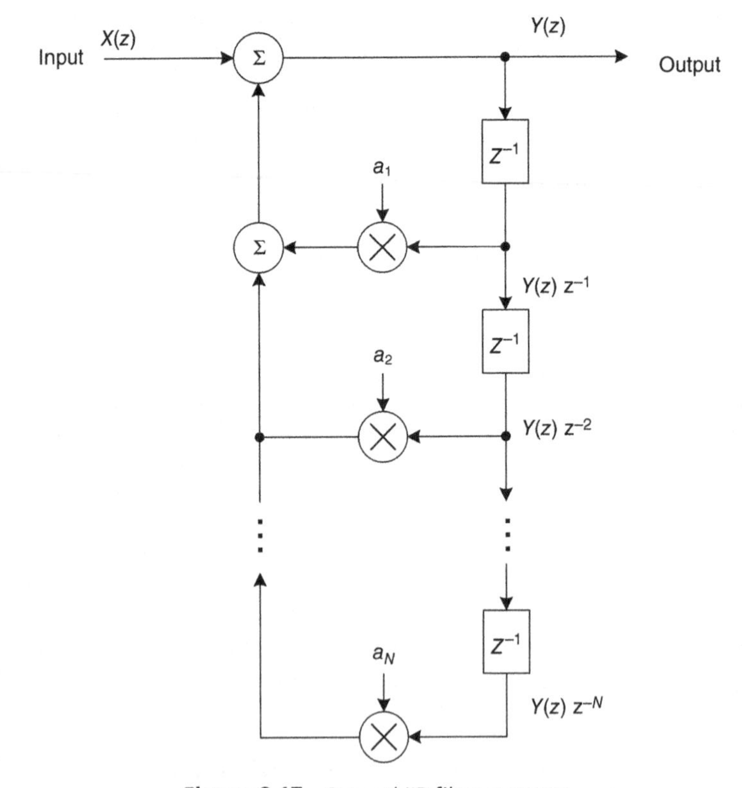

Figure 8.17. General IIR filter structure.

resulting in the transfer function

$$\frac{Y(z)}{X(z)} = \frac{1}{1 - a_1 z^{-1} - a_2 z^{-2} - \cdots - a_N z^{-N}} \tag{8.26}$$

To illustrate a critical difference between FIR and IIR filters, we inspect the transfer functions of the FIR and IIR filters from Equations 8.19 and 8.26. Multiplying numerator and denominator of the FIR transfer function by z^{N-1} yields Equation 8.27 which has $N-1$ poles at the origin and N–1 zeros $z_1, z_2, \ldots, z_{N-1}$. Since poles and zeros at the origin do not affect the magnitude of the frequency response (see footnote 8), the poles in the denominator of Equation 8.27 can be ignored and we see that the magnitude of the frequency response of the FIR filter is due to zeros only:

$$\begin{aligned} H_{\text{FIR}}(z) &= \sum_{n=0}^{N-1} b_n z^{-n} = \frac{1}{z^{N-1}} \sum_{n=0}^{N-1} b_n z^{N-1-n} \\ &= \frac{1}{z^{(N-1)}} \left(z - z_1\right) \left(z - z_2\right) \cdots \left(z - z_{(N-1)}\right) \end{aligned} \tag{8.27}$$

Similarly, multiplying numerator and denominator of the IIR transfer function of Equation 8.26 by z^N yields Equation 8.28, which has N zeros at the origin and N poles, $p_1, p_2, \ldots, p_N$.

$$\begin{aligned} H_{\text{IIR}}(z) &= z^N \frac{1}{z^N - a_1 z^{(N-1)} - a_2 z^{(N-2)} - \cdots - a_N} \\ &= z^N \frac{1}{\left(z - p_1\right)\left(z - p_2\right) \cdots \left(z - p_N\right)} \end{aligned} \tag{8.28}$$

Since IIR filters can implement poles, they can be used to implement classical all-pole analog filter responses such as Butterworth, Chebyshev, and Bessel as described in Chapter 7. The design of these digital filters is beyond the scope of this text, but the reader should be aware that digital filters are often designed based on analog prototypes using filter design software tools or Matlab.

8.3.3 Comparisons between FIR and IIR Filters

Some comparisons between FIR and IIR filters are summarized in Table 8.1.

TABLE 8.1. Some comparisons between FIR and IIR filters

	FIR	IIR
Stability	Unconditionally stable	Potentially unstable
Linear phase	Yes, if taps are symmetric	Generally not
Complexity	More complex	Less complex
Analog filter emulation	Typically not done	Commonly done
Absolute delay	Longer delay if taps are symmetric	Shorter delay

8.4 DESIGN OF SIMPLE AND PRACTICAL DIGITAL FILTERS

This section presents two digital filters that will satisfy many practical filtering requirements and can be easily designed using a spreadsheet program such as Microsoft Excel. These filters can be implemented using a minimal amount of hardware[10] or a few lines of fixed-point C code.

8.4.1 Averaging Lowpass FIR Filter

The structure in Figure 8.18 is called an *averaging filter* because the output is simply the average of N consecutive samples of the input sequence. This filter offers the following benefits:

- It can be implemented with no multipliers.
- It can be implemented using a few lines of fixed-point C code.
- It has nulls in its frequency response that can be used to eliminate signals at specific frequencies.
- It is linear phase and therefore desirable for signals consisting of multiple frequencies.

The filter uses the FIR structure of Figure 8.12 with all tap weights set to unity. If the number of taps is a power of 2, that is, 2^k, then the division by N at the filter output can be done by shifting the result of the summation right by k bit positions thereby eliminating the need for a multiplier.

The z-domain transfer function is given by Equation 8.19 with all tap weights set to unity resulting in:

$$H(z) = \frac{1}{N} \sum_{n=0}^{N-1} z^{-n} \tag{8.29}$$

[10]These filters can be implemented using bit shifts and additions instead of multipliers. This reduces hardware complexity and power consumption.

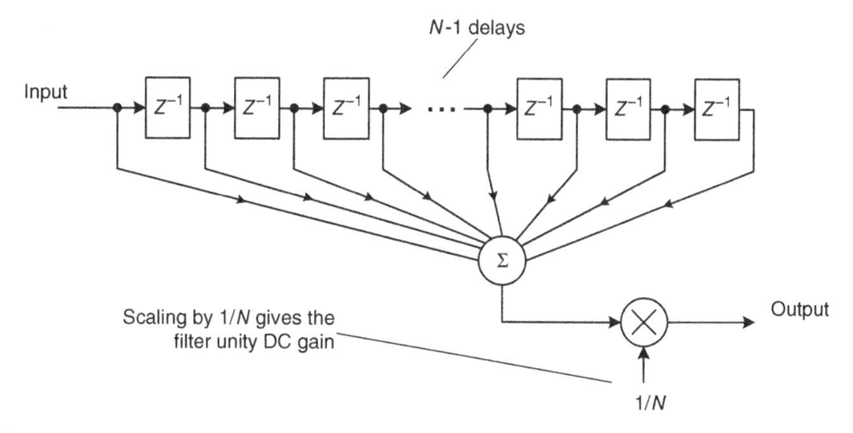

Figure 8.18. Averaging lowpass FIR filter. This practical filter simply sums all the samples in the register and divides the result by *N*, resulting in their average.

The summation operation of Equation 8.29 can be eliminated to produce a more convenient transfer function. First we recognize that Equation 8.29 is the geometric series [1]

$$H(z) = \frac{1}{N}\sum_{n=0}^{N-1}\left(z^{-1}\right)^{n} \tag{8.30}$$

and represent it using the formula for the sum[11]

$$H(z) = \frac{1}{N}\cdot\frac{1-z^{-N}}{1-z^{-1}} \tag{8.31}$$

In the preceding sections we computed the frequency response of z-domain transfer functions by making the substitution $z = e^{sT_S}$ and using Matlab to compute the complex quantities. We can simplify the design and analysis of the averaging lowpass filter by deriving an equation that allows us to compute the magnitude of the frequency response without having to manipulate complex values.

Substituting $z = e^{sT_S}$ into Equation 8.31 gives

$$H(s) = \frac{1}{N}\cdot\frac{1-e^{-sNT_S}}{1-e^{-sT_S}} \tag{8.32}$$

[11]This equation is the basis for a filter structure called the cascaded integrator comb (CIC) lowpass filter which is frequently used for reducing the sample rate of signals after lowpass filtering. This simple and elegant structure is described in [3] and uses even less hardware than Figure 8.18!

Using the identity $e^{sT_S} = \cos\left(\omega T_S\right) + j\sin\left(\omega T_S\right)$ we rewrite Equation 8.32 as

$$H(\omega) = \frac{1}{N} \cdot \frac{1 - \cos\left(\omega N T_S\right) + j\sin\left(\omega N T_S\right)}{1 - \cos\left(\omega T_S\right) + j\sin\left(\omega T_S\right)} \tag{8.33}$$

The magnitude of the frequency response is

$$|H(\omega)| = \frac{1}{N} \cdot \frac{\sqrt{\left(1 - \cos\left(\omega N T_S\right)\right)^2 + \sin^2\left(\omega N T_S\right)}}{\sqrt{\left(1 - \cos\left(\omega T_S\right)\right)^2 + \sin^2\left(\omega T_S\right)}} \tag{8.34}$$

The identity $\sin^2(\phi) + \cos^2(\phi) = 1$ allows us to write Equation 8.34 as

$$|H(\omega)| = \frac{1}{N} \cdot \frac{\sqrt{1 - \cos\left(\omega N T_S\right)}}{\sqrt{1 - \cos\left(\omega T_S\right)}} \tag{8.35}$$

And the identity $\sin^2(\phi) = 1/2 \cdot (1 - \cos(2\phi))$ allows us to write Equation 8.35 as

$$|H(\omega)| = \frac{1}{N} \cdot \left| \frac{\sin\left(\frac{\omega N T_S}{2}\right)}{\sin\left(\frac{\omega T_S}{2}\right)} \right| \tag{8.36}$$

Replacing ω by $2\pi f$ in Equation 8.36 gives

$$|H(f)| = \frac{1}{N} \cdot \left| \frac{\sin\left(\pi N \frac{f}{F_S}\right)}{\sin\left(\pi \frac{f}{F_S}\right)} \right| \tag{8.37}$$

where the absolute value operator keeps the result positive so it can be plotted in decibels.

Inspection of the numerator of Equation 8.37 shows that this filter has frequency response nulls at all nonzero multiples of F_S/N. By selecting the sample rate and filter length, these nulls can be strategically placed to eliminate signals at specific frequencies as shown in Problem 8.5. Microsoft Excel was used to plot Figure 8.19 which is the frequency response of an averaging FIR with $N = 8$ and $F_S = 10$ kHz.

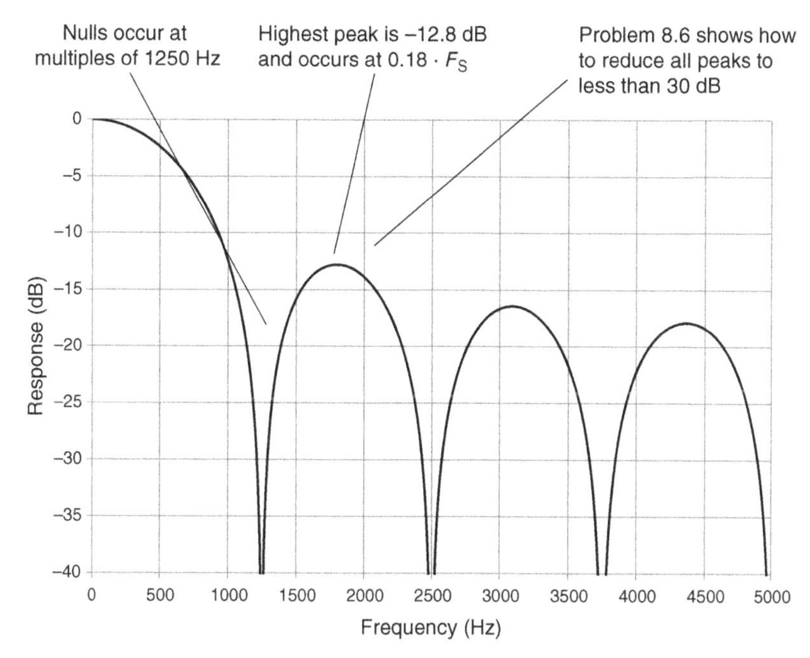

Figure 8.19. Frequency response for averaging lowpass FIR filter with $N = 8$ and $F_S = 10$ kHz from Equation 8.37. The exact frequency and location of the highest peak depend on N.

Better stopband rejection can be achieved by cascading two or more averaging FIR filters. Problem 8.6 shows that cascading a seven-tap and five-tap filter results in greater than 30 dB stopband rejection.

The averaging lowpass filter can be implemented in firmware using a circular buffer and a static variable that represents the sum of the values in the buffer. The filter is iterated by subtracting the oldest register value from the sum, adding the current input, and then placing the input in the buffer. This makes the computational requirements of the filter independent of the register length. Figure 8.20 shows how to implement the averaging lowpass filter using only bit shifts and additions in C code.

8.4.2 Lowpass FIR/IIR Filter

The lowpass filter shown in Figure 8.21 offers the following benefits:

- It is simple to specify because its response approximates the lowpass RC filter.[12]

[12]Design of the lowpass RC filter is discussed in Chapter 7.

```
1   #define K    3              // The register length must be a power of two to avoid multiplies.
2   #define N    (int) (1<<K)   // Register length is 2^k.
3
4   int AveragingFilter (int input) {
5
6       static int   circ_buf[N];      // The circular buffer is the filter register.
7       static long  sum;              // Sum of elements in the filter register.
8       static int   circ_buf_ptr;     // Index for the circular buffer.
9
10      sum -= circ_buf[circ_buf_ptr];       // Subtract the oldest value from the sum.
11      circ_buf[circ_buf_ptr] = input;      // Place the input in the filter register.
12      sum += circ_buf[circ_buf_ptr++];     // Add the newest value to the sum.
13      circ_buf_ptr %= N;                   // Increment the buffer keeping it in the range 0 to N-1;
14
15      return ((int)(sum >> K));            // Scale by 2^-K to maintain unity gain.
16  }
```

Figure 8.20. C code listing for the lowpass averaging filter.

- It can be implemented with no multipliers and therefore minimal FPGA logic.
- It can be implemented using a few lines of fixed-point C code.
- It can be easily converted to a highpass filter (Problem 8.9).

Since the filter approximates the response of the lowpass RC filter, it will attenuate by 3 dB at the cutoff frequency, and the response will fall off at 6 dB per octave or equivalently 20 dB per decade. This allows us to use the design techniques of Chapter 7 to specify the cutoff frequency for a given filtering requirement. Our next task is to develop a design equation that relates the coefficient α from Figure 8.21 and the sample rate to the cutoff frequency. As before, we begin by computing the z-domain transfer function.

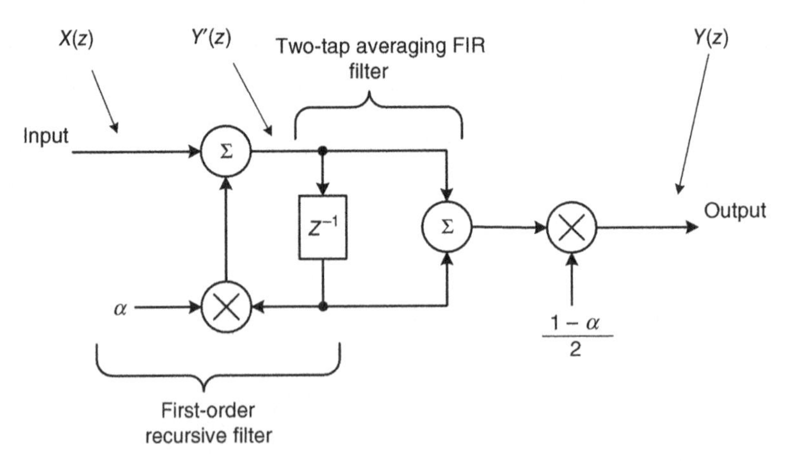

Figure 8.21. Lowpass FIR/IIR filter. The left side of the filter is a first-order IIR filter. The right side is an averaging FIR. The overall filter response approximates the lowpass RC filter.

Inspection of Figure 8.21 shows that this filter has the same structure as the filter shown in Figure 8.6 with the z-domain transfer function shown in Equation 8.14. With $a_1 = \alpha$, $b_0 = 1$, $b_1 = 1$, and all other coefficients zero, Equation 8.14 becomes

$$H(z) = \frac{1-\alpha}{2} \cdot \frac{1+z^{-1}}{1-\alpha z^{-1}} = \frac{1-\alpha}{2} \cdot \frac{z+1}{z-\alpha} \tag{8.38}$$

In the previous sections we computed the exact frequency response for z-domain transfer functions by substituting $z = e^{sT_S}$ for z. To illustrate the relationship between this filter and the lowpass RC filter we approximate z using the first two terms of its power series representation[13]

$$z = e^{sT_S} \cong 1 + sT_S \tag{8.39}$$

which is valid when the product of sT_S is small or equivalently when the frequency is low relative to the sample rate.[14]

Substituting Equation 8.39 into Equation 8.38 gives the approximation

$$H(s) \cong \frac{1-\alpha}{2} \cdot \frac{s + \frac{2}{T_S}}{s + \frac{1-\alpha}{T_S}} \tag{8.40}$$

Since $2/T_S$ is large compared to values of s that will be used to compute the frequency response, the numerator of the fraction on the right-hand side of Equation 8.40 can be replaced by $2/T_S$ and Equation 8.40 becomes

$$H(s) \cong \frac{1-\alpha}{T_S} \cdot \frac{1}{s + \frac{1-\alpha}{T_S}} \tag{8.41}$$

or

$$H(s) \cong \frac{1}{\tau} \cdot \frac{1}{s + \frac{1}{\tau}} \tag{8.42}$$

[13]This approximation is called the *method of backward differences*. It gives a simple mapping between the s and z domains that is accurate at low frequencies.

[14]We can also say that this approximation is valid provided the sample rate is much higher than the bandwidth of the filter.

where

$$\tau = \frac{T_S}{1 - \alpha} \tag{8.43}$$

We recognize Equation 8.42 as the s-domain transfer function of the lowpass RC filter as discussed in Chapter 7, where the cutoff frequency was shown to be

$$F_C = \frac{1}{2\pi\tau} = \frac{1}{2\pi RC} \tag{8.44}$$

Substituting Equation 8.43 into Equation 8.44 and solving for α gives the equation relating α to the sample rate and cutoff frequency:

$$\alpha = 1 - 2\pi\frac{F_C}{F_S} \tag{8.45}$$

Example 8.4. Specify the sample rate and feedback coefficient for the lowpass FIR/IIR filter of Figure 8.21 so its cutoff frequency is approximately 170 Hz and it provides at least 30 dB attenuation at 8 kHz. Implement the filter without using multipliers.

Solution. Using the techniques of Chapter 7 we note that the number of octaves between 170 and 8 kHz is

$$\text{Octaves} = \frac{\ln\left(\frac{8000}{170}\right)}{\ln(2)} = 5.56 \tag{8.46}$$

Since the lowpass RC filter rolls off at 6 dB per octave, it would provide $5.56 \times 6 = 33.4$ dB attenuation at 8 kHz. Since we require only 30 dB attenuation at 8 kHz, the lowpass RC filter should suffice for this requirement.

Since the frequency response of any digital filter is periodic, we expect the filter to provide roll-off up to half the sample rate. Above this frequency, the filter response characteristic begins to rise again. Therefore, we set the half sample rate to 8 kHz resulting in a sample rate of 16 kHz. From Equation 8.45

$$\alpha = 1 - 2\pi\frac{F_C}{F_S} = 1 - 2\pi \times \frac{170}{16{,}000} = 0.9332 \tag{8.47}$$

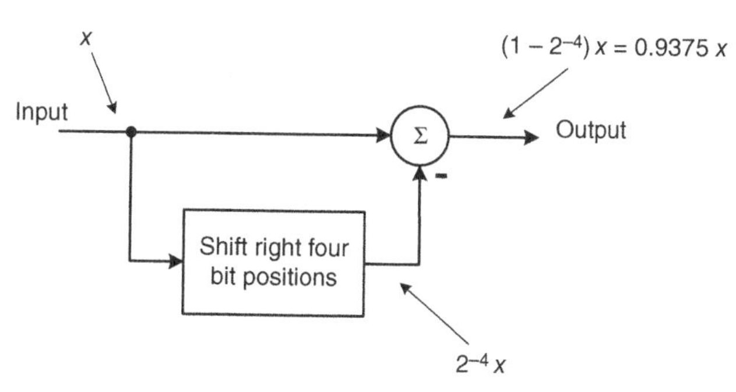

Figure 8.22. Multiplication by a factor of $1 - 2^{-k}$ can be done using bit shifting and subtraction instead of multipliers.

It is advantageous to use values of α that can be represented as $1 - 2^{-k}$ because the multiplication by α in Figure 8.21 can be done with bit shifts and subtraction as shown in Figure 8.22. Therefore, we approximate our computed value of α with $1 - 2^{-4}$ or 0.9375.

The filter of Figure 8.21 uses output scaling of $(1 - \alpha)/2$ to maintain unity DC gain. This can be implemented by a shift operation because

$$\frac{1-\alpha}{2} = \frac{1-\left(1-2^{-k}\right)}{2} = 2^{-(k+1)} \tag{8.48}$$

We used the approximation $z = e^{sT_S} \cong 1 + sT_S$ to design the filter, but now wish to check its exact frequency response. Using a procedure very similar to that shown in Equation 8.32 to Equation 8.37 (see Problem 8.8) we compute the frequency response of the lowpass FIR/IIR filter as

$$|H(f)| = \left| \frac{(1-\alpha)\cos\left(\pi \frac{f}{F_S}\right)}{\sqrt{1+\alpha^2 - 2\alpha\cos\left(2\pi \frac{f}{F_S}\right)}} \right| \tag{8.49}$$

With the feedback coefficient set to 0.9375 the frequency response of the filter is plotted in Figure 8.23. Figure 8.24 shows the filter of Example 8.4 implemented using only bit shifts and additions. Figure 8.25 shows how to implement the filter of Example 8.4 using only bit shifts and additions in C code.

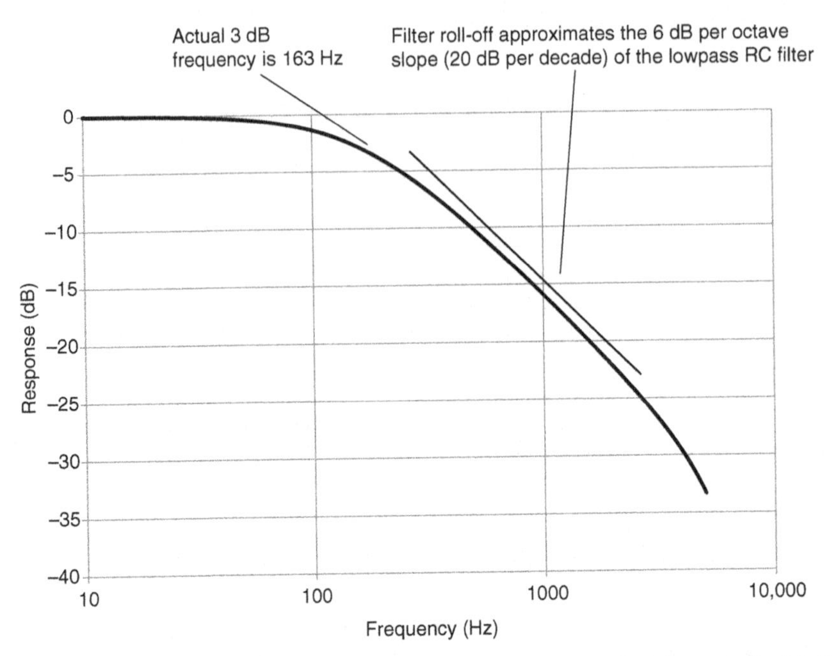

Figure 8.23. Frequency response for lowpass FIR/IIR filter specified in Example 8.4. The sample rate of 16 kHz and feedback factor of 0.9375 satisfy the design requirements.

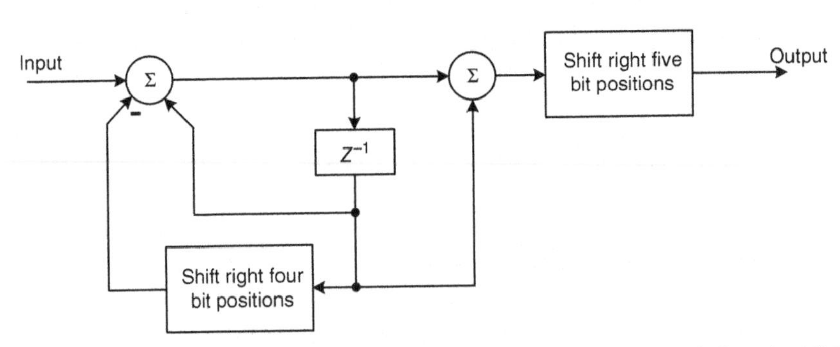

Figure 8.24. The lowpass FIR/IIR filter can be implemented using only shift and addition operations. The DC gain of this filter is unity.

```
1   #define FILTER_SHIFT 4              // Parameter k
2
3   int LowpassFirIirFilter (int input) {
4
5       static long     filter_reg;         // Register is 32 bits.
6       long            filter_reg_store;   // Temporary storage.
7
8       filter_reg_store = filter_reg;      // Temporarily store the last register value.
9
10      // Update register with the current input sample.
11      filter_reg = filter_reg - (filter_reg >> FILTER_SHIFT) + input;
12
13      // Update the FIR section and scale output.
14      return((int)((filter_reg + filter_reg_store) >> (FILTER_SHIFT + 1)));
15  }
```

Figure 8.25. C code listing for the filter of Example 8.4. Note that the register is 32 bits while the input and output are 16 bits wide. See Problem 8.10.

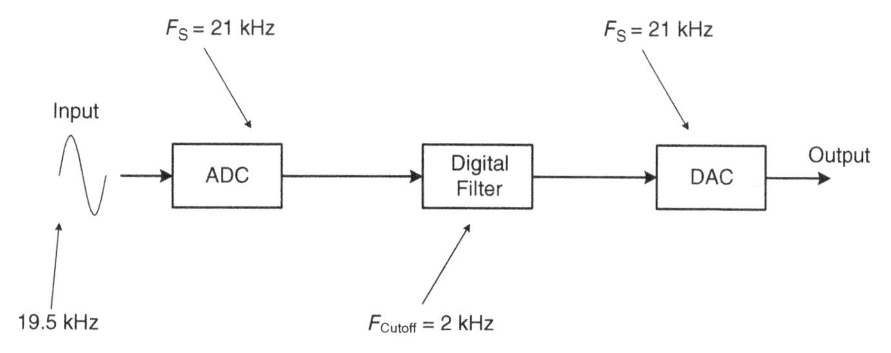

Figure 8.26. System for Problem 8.1.

PROBLEMS

8.1 The ADC in Figure 8.26 samples a 19.5 kHz sinusoid at 21 kHz. The digital lowpass filter passes signals below 2 kHz and rejects signals above 2 kHz.

(a) If a signal at 19.5 kHz were fed to the system would it be passed or rejected by the filter?

(b) What frequency would be seen at the DAC output?

(c) If a brick wall filter with cutoff at the Nyquist frequency were placed in front of the ADC, what would be seen at the DAC output?

8.2 The anti-aliasing filter in Figure 8.27 has 3 dB cutoff frequency 100 kHz and a two-pole response, meaning it rolls off at 12 dB/octave. Determine the minimum sample rate so that all aliased components are attenuated by at least 36 dB.

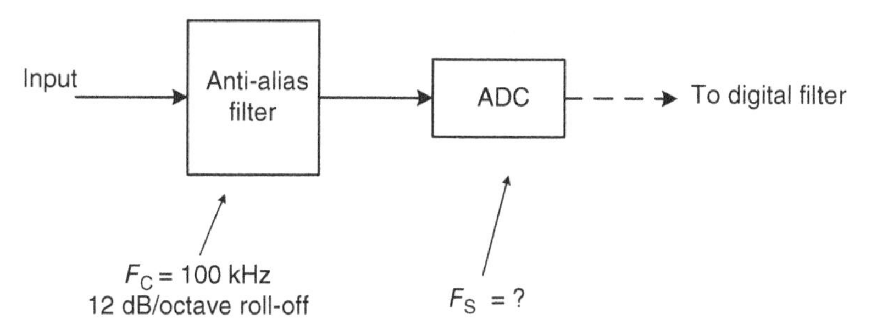

Figure 8.27. System for Problem 8.2.

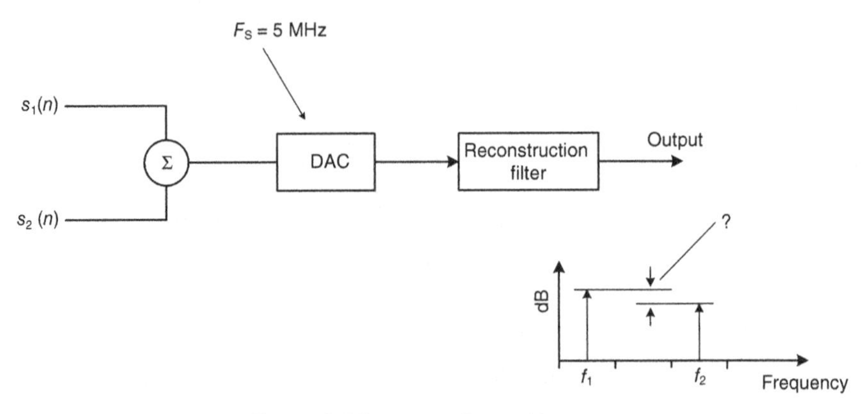

Figure 8.28. System for Problem 8.3.

8.3 The input to the DAC in Figure 8.28 consists of two sampled sinusoids of equal amplitude. The reconstruction filter attenuates both frequencies equally:

$$s_1(n) = \sin\left(2\pi\frac{n}{50}\right)$$

and

$$s_2(n) = \sin\left(2\pi\frac{n}{5}\right)$$

The sample rate for the DAC is 5 MHz.

(a) What are the two frequencies seen at the DAC output?

(b) What is the difference between the two sinusoids at the output in decibels?

8.4 The digital filter in Figure 8.29 is called a *unity-gain resonator* and is sometimes used for digital audio equalizers. The sample rate for the filter is 44.1 kHz.

(a) Determine the z-domain transfer function for the structure.

(b) Determine the DC gain by inspection of the transfer function.

(c) Plot the frequency response for the structure.

8.5 Design an averaging lowpass FIR filter that will pass a desired DC signal but will reject 60 Hz hum from the power line as well as all harmonics of 60 Hz. Choose the closest sampling rate to 500 Hz and select the value of N so that 60 Hz and its harmonics are located at the filter nulls. What is the attenuation of the first harmonic if the hum is 63 Hz instead of 60 Hz?

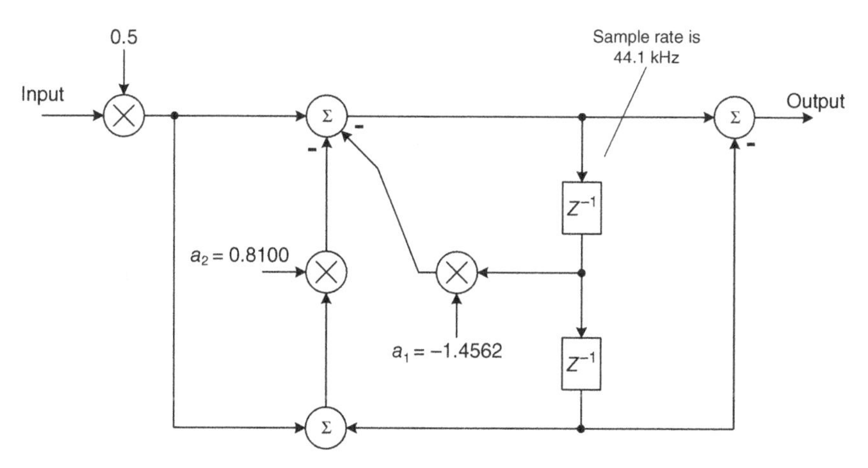

Figure 8.29. Unity-gain resonator for Problem 8.4.

8.6 The output of an $N = 5$ averaging lowpass filter is connected to the input of an $N = 7$ averaging lowpass filter. Plot the frequency response if the sample rate is 100 kHz. Comment on why this filter would be used instead of a single averaging FIR filter.

8.7 Show that the ratio of the cutoff frequency of the lowpass FIR/IIR filter from Section 8.4.2 to the sample rate is

$$\frac{F_C}{F_S} = \frac{1}{2\pi} 2^{-k} \tag{8.50}$$

For a given sample rate how is the cutoff frequency affected by increasing k by 1?

8.8 Derive Equation 8.49 for the exact magnitude response of the lowpass FIR/IIR filter of Figure 8.21.

8.9 The lowpass FIR/IIR can be converted to a highpass filter by subtracting its output from the filter input as shown in Figure 8.30a.

(a) Show that the z-domain transfer function for the highpass filter is

$$H(z) = 1 - \frac{1 - \alpha}{2} \cdot \frac{z + 1}{z - \alpha} \tag{8.51}$$

(b) Substitute $z = 1 + sT_S$ into your transfer function and show that the frequency response approximates the response of the highpass RC filter shown in Figure 8.30b.

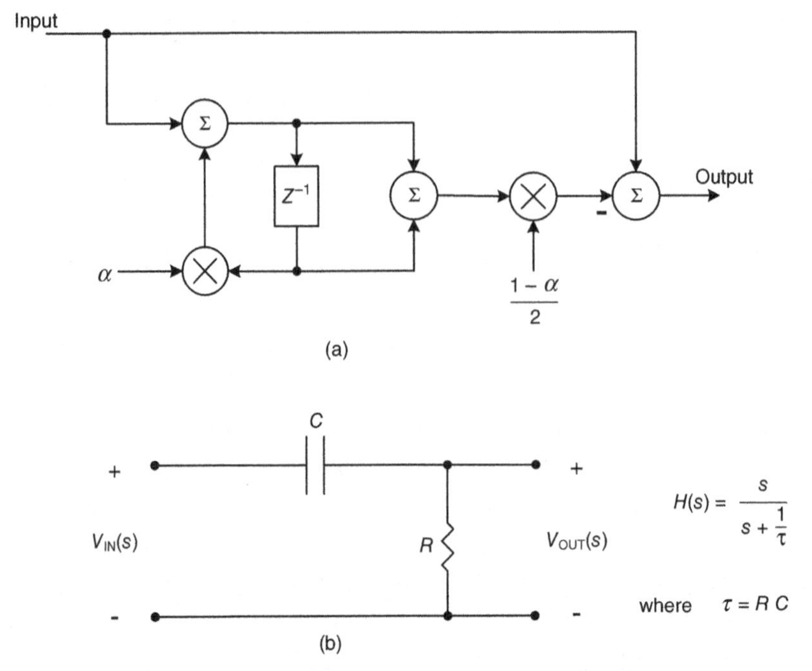

Figure 8.30. The highpass filter shown in (a) is based on the lowpass FIR/IIR filter. This filter approximates the highpass RC filter shown in (b).

(c) Show that the approximate highpass 3 dB frequency for the digital filter is

$$F_{3\mathrm{dB}} \cong \frac{1}{2\pi} \cdot \frac{1-\alpha}{T_S} \tag{8.52}$$

Hint: You will need to approximate $\alpha = 1$ at one point in the algebra.

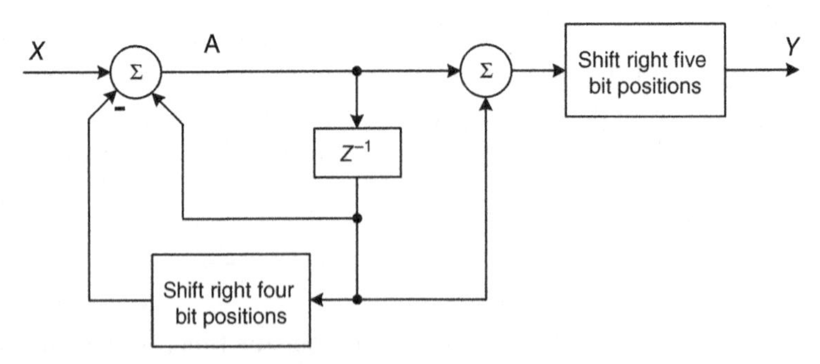

Figure 8.31. Lowpass FIR/IIR filter. When implementing this filter you must account for the DC gain from the input to point A which is greater than unity.

8.10 Compute the transfer function between point A and the input, $H(z) = A(z)/X(z)$, for the lowpass FIR/IIR shown in Figure 8.31. Write a formula for the DC gain of the transfer function. Comment on why the C code in Figure 8.25 uses a 32-bit integer for the register.

REFERENCES

1. Wikipedia. *Geometric Series.* Available: http://en.wikipedia.org/wiki/Geometric_series (Accessed 11-30-2013.)
2. F. J. Harris, *Multirate Signal Processing*, Prentice Hall, 2004.
3. Wikipedia. *Cascade Integrator-Comb Filter.* Available: http://en.wikipedia.org/wiki/CIC_filter (Accessed 11-30-2013.)

9

HOW TO WORK WITH RF SIGNALS

Whether you work with digital circuitry or radio systems, an understanding of radio frequency (RF) signals will enable you to design circuitry that is reliable and manufacturable. RF circuitry is sometimes taught as an elective course only, and you may have missed it. If this applies to you, don't be intimidated by the strange plots, the number of equations, or the length of this chapter. You will pick up valuable insights and tools—even if it's your first venture into this fascinating area.

Engineers sometimes make the mistake of designing low-speed (<50 MHz) circuitry without regard to the physical connections between components. For example, given a printed circuit trace between two devices as shown in Figure 9.1, the following assumptions are often made:

1. The voltage at A is always identical to that at B.
2. Current injected at A is always identical to that at B.
3. The impedance measured at A is identical to the impedance measured at B.

Ten Essential Skills for Electrical Engineers, First Edition. Barry L. Dorr.

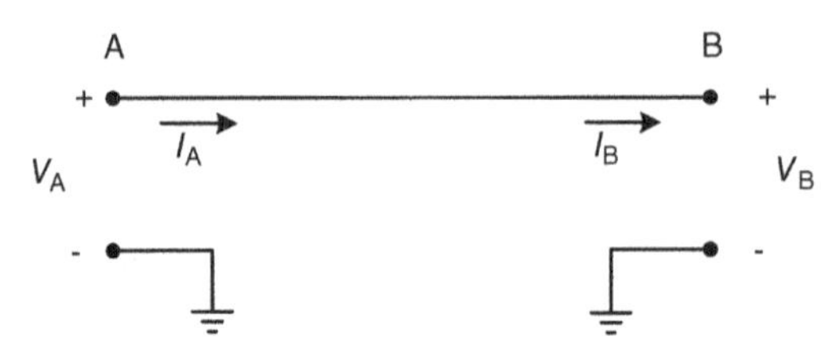

Figure 9.1. Connection between two points of a circuit. This notion is useful at low frequencies but doesn't account for high-frequency effects.

If the physical distance between devices is small compared to the wavelengths[1] of the signals involved, these are good assumptions.[2] But as frequency increases, these assumptions become less valid, and circuitry doesn't always work as expected. This can result in op-amps that oscillate, digital devices that don't work properly, or simply "weird and unexplainable" behavior. Recalling that commercial op-amps often have significant gain at high frequency and that fast transitions from digital gates contain high-frequency content, high-frequency behavior of printed circuit traces should be considered for any design.

Engineers, when faced with "weird and unexplainable" behavior, sometimes abandon the fundamentals in favor of "black magic." After all, a circuit may work when the engineer touches his finger to the trace, and then fail when he removes it. Or he may find that a digital circuit fails when a circuit board is mounted in a chassis, but works when the mounting screws for the board are loosened. In reality, issues such as these can always be traced back to the fundamentals. The best RF engineers learn from experienced technicians and engineers by critically examining the "black magic," relating it to the fundamentals and then including both practice and theory into their skill sets. This chapter will introduce you to some of the fundamentals.

The material up to and including Section 9.4 is pertinent to both digital and RF engineers. The remainder of the chapter is targeted primarily toward RF engineers. We begin with a discussion of energy transfer where we see that energy travels in fields, not conductors. With this understanding we then examine the phenomena of signal reflections and problems caused by reflections in high-speed digital circuits. Finally, we review use of the Smith chart for working with transmission lines and impedances.

[1]Wavelength is inversely proportional to frequency.

[2]Especially when drawing a schematic diagram!

9.1 ENERGY TRANSFER

To work effectively with RF signals it is necessary to understand that energy travels in fields and not conductors. Figure 9.2 shows a schematic diagram of a DC voltage source connected to a resistive load. We know that energy is being transferred from the power supply to the load, and the schematic suggests that the energy is transferred through the wires. This notion is useful because it allows us to represent circuits using schematic diagrams. But it does not explain why we see signal reflections on printed circuit board traces or why one trace on a printed circuit board can couple a signal to another. It certainly doesn't explain why it's possible to transfer power from a power amplifier into a transmitting antenna!

A good way to remind ourselves that energy travels in fields is to consider a simple use of the Poynting vector defined as

$$\mathbf{S} = \mathbf{E} \times \mathbf{H} \tag{9.1}$$

where

S is the energy density vector in watts per square meter
E is the electric field vector
H is the magnetic field vector
$\times$ denotes the cross product.

This relationship is valid at all frequencies including DC. Operations with the Poynting vector frequently involve complex vector calculus, but our application is a simple use of the right-hand rule. In Figure 9.3a, consider the case of a DC source supplying power to a resistive load. Figure 9.3b is a cross-sectional view of the area near the upper conductor and shows the electric and magnetic fields as the current flows toward the reader. The right-hand rule

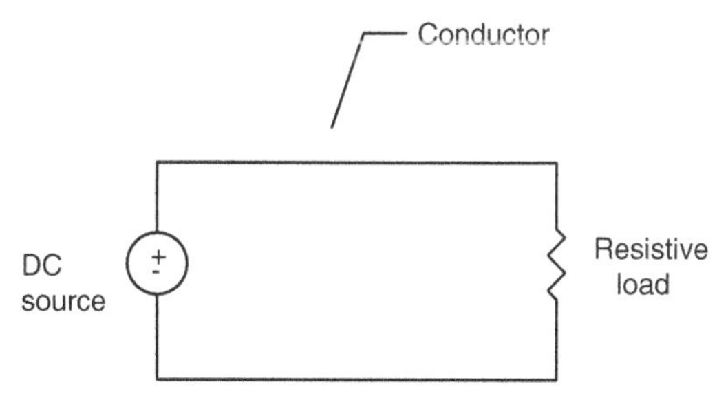

Figure 9.2. Power supply delivering power to a load. The drawing suggests that energy is transferred in the conductor, but it is actually transferred in the fields surrounding the conductor.

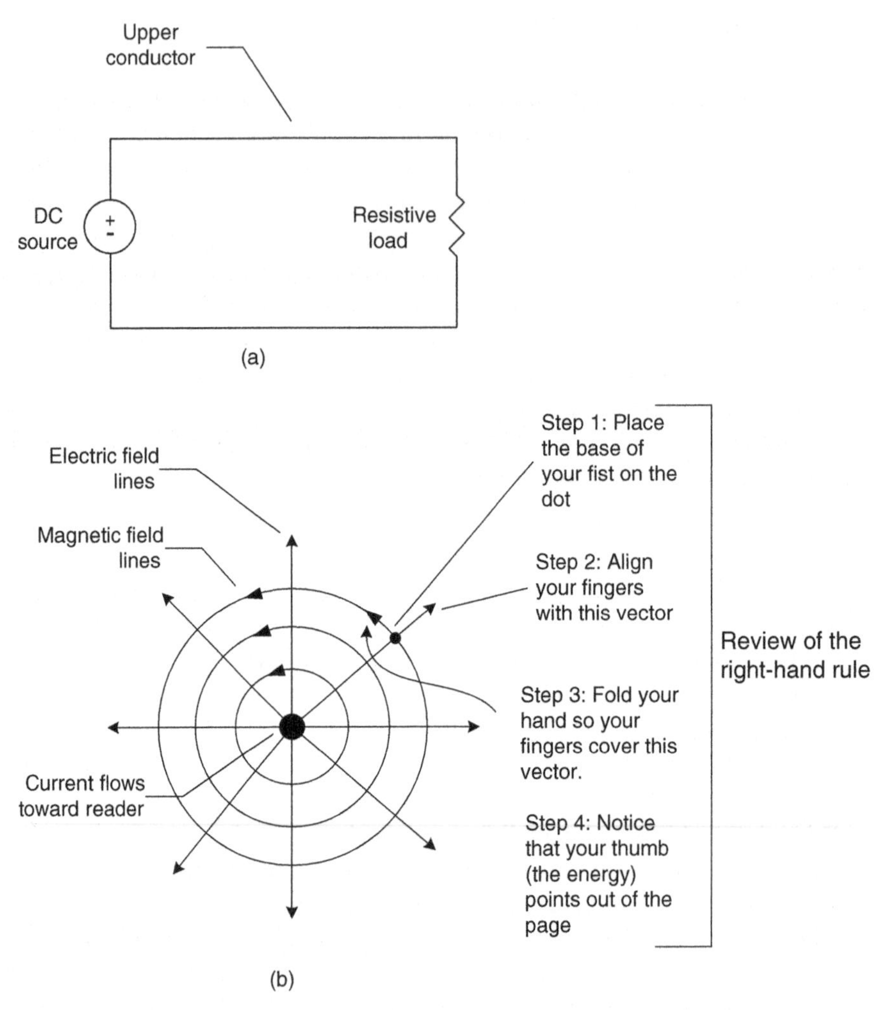

Figure 9.3. (a) DC circuit. (b) Cross-sectional view of conductor showing electric and magnetic fields near the upper conductor. The current and energy transfer is toward the reader.

for the field shows that the energy is flowing toward the reader. If the load were shorted to ground then we intuitively know there is no energy transfer. Equation 9.1 predicts this because the **E** field is zero. Similarly disconnecting the load sets the **H** field to zero and again there is no energy transfer.

It is particularly important to keep this concept in mind when working with printed circuit boards where conductors are in close proximity. With time-varying signals, fields are time varying and a voltage on one conductor can induce a voltage on a nearby conductor due to the capacitance between them. Similarly a conductor can induce a current in a nearby conductor by transformer action. Design rules for printed circuit board design are largely derived from the fact that energy travels exclusively in fields.

9.2 SIGNAL REFLECTIONS

We can get a physical feel for signal reflections by considering a flexible membrane placed in a tank as shown in Figure 9.4. The left side of the tank is filled with water and the right side is filled with a viscous liquid such as honey. If we create a wave at the left end of the tank it will travel toward the membrane (a). But when it reaches the membrane (b) the honey won't move as easily as the water. This causes a portion of the forward wave to be reflected back toward its source (c). It also causes a wave on the right side of the tank.

The wave analogy above is analogous to the situation where a source drives a transmission line such as a coaxial cable terminated in a mismatched load as shown in Figure 9.5. The transmission line has characteristic impedance equal to $\sqrt{L/C}$ where L and C are per unit length values of inductance and capacitance, respectively. A common value for the impedance of coaxial cables is 50 Ω. Though the cable is described by resistive impedance it is lossless and does not dissipate energy.[3] Instead, energy traveling through the cable encounters a 50 Ω impedance as it moves. Intuitively we can visualize the cable as a "bucket brigade" where energy is passed from bucket to bucket but not spilled. If the cable is terminated in its characteristic impedance then the energy is dissipated as heat. If not, a portion of it is reflected and travels back to the source.

By considering what happens when the forward wave reaches the load we can write several descriptive equations:

1. The total voltage at the load, V_L, is equal to the sum of the forward and reflected voltages.
2. The total current at the load, I_L, is equal to the forward current minus the reflected current.
3. The voltage–current relationship at the load must follow Ohm's law.
4. The forward and reflected waves have a voltage–current ratio equal to the characteristic impedance of the cable.

We define the complex reflection coefficient, K_L, as the fraction of the voltage reflected at the boundary between the cable and the load. It is also

[3] In Section 9.4 it is shown that practical cables have a small amount of loss.

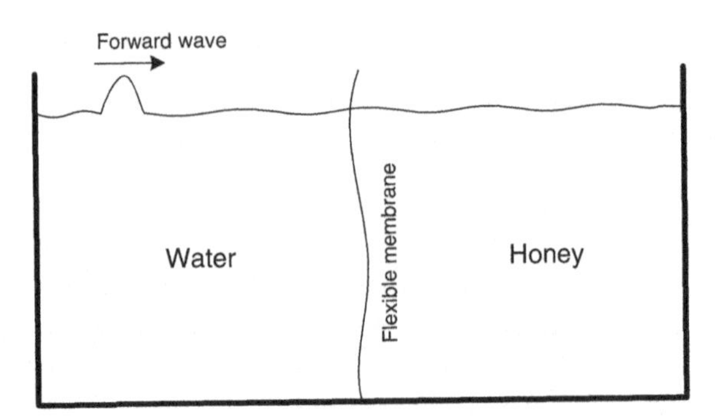

(a) The forward wave travels toward the membrane

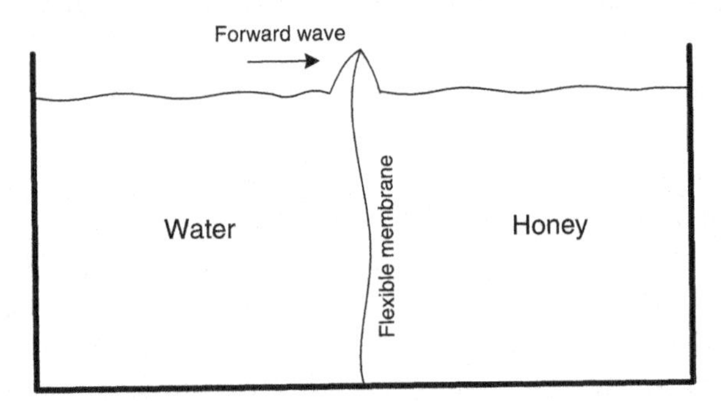

(b) The forward wave reaches the membrane and transfers some of its energy to the honey

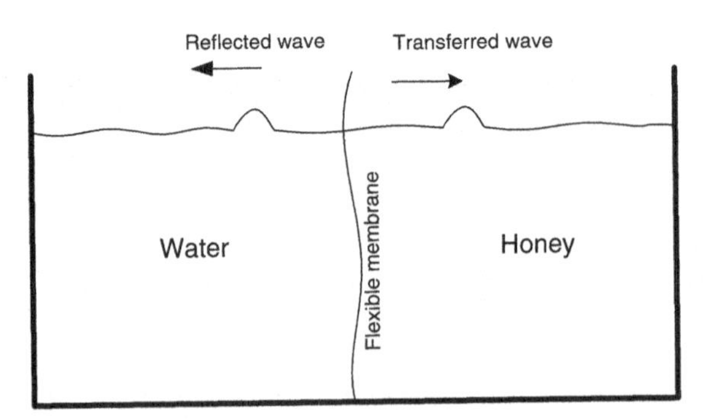

(c) The transferred wave continues through the honey while the reflected wave returns to the source

Figure 9.4. Wave reflections in a tank containing different liquids. Similarly, when an electromagnetic signal encounters a change in its propagation medium some of the energy is transferred and some is reflected.

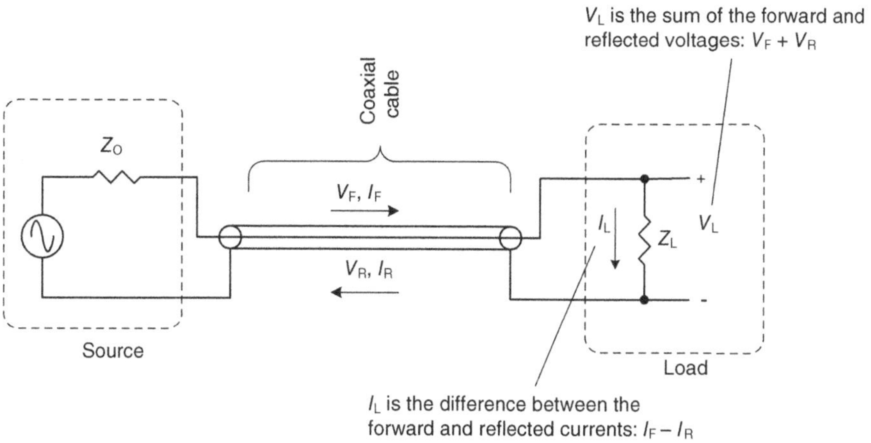

Figure 9.5. Sinusoidal source driving a mismatched load through a coaxial cable. Ohm's law is always satisfied at the load and the forward and reflected waves are always related by the impedance of the cable.

equal to the fraction of the current reflected at the load. We will see there is a direct relationship between the reflection coefficient and the load impedance. This relationship will be exploited when we discuss the Smith chart later in Sections 9.5, 9.6, and 9.7. We relate the reflection coefficient to the load impedance using the following steps:

From (1) above

$$V_L = V_F + V_R = V_F + K_L V_F = V_F\left(1 + K_L\right) \tag{9.2}$$

where V_L is the voltage at the load, V_F is the forward voltage, V_R is the reverse voltage, and K_L is the reflection coefficient.

From (2) above

$$I_L = I_F - I_R = I_F\left(1 - K_L\right) \tag{9.3}$$

where I_L is the current at the load, I_F is the forward current, I_R is the reverse current, and K_L is the reflection coefficient.

Combining Equations 9.2 and 9.3 gives

$$Z_L = \frac{V_L}{I_L} = \frac{V_F}{I_F}\frac{\left(1 + K_L\right)}{\left(1 - K_L\right)} = Z_O\frac{\left(1 + K_L\right)}{\left(1 - K_L\right)} \tag{9.4}$$

where Z_O is the characteristic impedance of the coaxial cable.

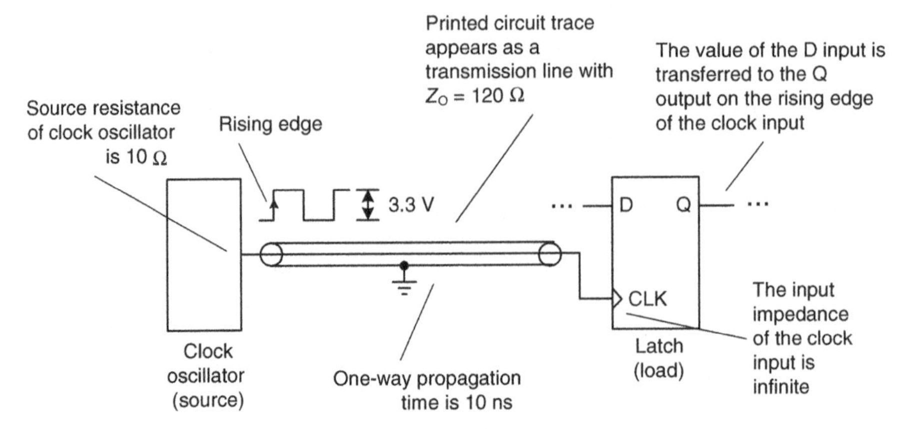

Figure 9.6. The clock oscillator feeds the clock input of a latch through a printed circuit board trace. If reflections are not considered, the circuit might not work as designed.

Solving Equation 9.4 for K_L gives

$$K_L = \frac{(Z_L - Z_O)}{(Z_L + Z_O)} \tag{9.5}$$

9.3 EFFECT OF SIGNAL REFLECTIONS ON DIGITAL SIGNALS

If signal reflections are not taken into account, digital circuits may not work as expected. Consider the case of a clock signal fed from a digital clock oscillator (source) to the clock input of a latch[4] (load) as shown in Figure 9.6. When the circuit is operating correctly, each rising edge of the clock signal causes a single rising edge at the latch's clock input, thereby causing the signal at the D input to be transferred to the Q output. For this example, the output impedance of the oscillator is 10 Ω, and the impedance of the clock input is infinite. The printed circuit board trace between the source and load appears as a transmission line with characteristic impedance 120 Ω. We will see that signal reflections can cause the latch to be clocked multiple times for a single rising edge of the clock signal thereby causing incorrect operation.

When the oscillator produces a rising edge, the voltage at its output is found by voltage division

$$V_F = 3.3 \times \frac{Z_O}{Z_O + R_S} = 3.3 \times \frac{120}{120 + 10} = 3.05 \text{ V} \tag{9.6}$$

[4]Also referred to as a D flip-flop.

and a signal, V_F, begins propagating through the printed circuit trace toward the load. The voltage of V_F is 3.05 V. When V_F reaches the load a reflection occurs. The reflection coefficient of the load is found using Equation 9.5:

$$K_L = \frac{(Z_L - Z_O)}{(Z_L + Z_O)} = \frac{(\infty - 120)}{(\infty + 120)} = 1 \tag{9.7}$$

The voltage at the load is found using Equation 9.2:

$$V_L = V_F(1 + K_L) = 3.05(1 + 1) = 6.1 \text{ V} \tag{9.8}$$

The reflected signal, V_R, travels toward the source. From the definition of the reflection coefficient, the voltage of V_R is

$$V_R = V_F K_L = 3.05 \times 1 = 3.05 \text{ V} \tag{9.9}$$

When the reflected signal, V_R, reaches the source, another reflection will occur. Equation 9.5 gives the source reflection coefficient as

$$K_S = \frac{(Z_s - Z_O)}{(Z_s + Z_O)} = \frac{(10 - 120)}{(10 + 120)} = -0.846 \tag{9.10}$$

The voltage at the source is the input voltage of 3.05 V plus the voltage caused by the reflection of signal V_R from the load:

$$V_S = V_F + V_R(1 + K_S) = 3.05 + 3.05(1 + (-0.846)) = 3.52 \text{ V} \tag{9.11}$$

and the voltage of the reflected signal is

$$V_{F2} = V_R K_S = 3.05(-0.846) = -2.58 \text{ V} \tag{9.12}$$

When V_{F2} reaches the load, another reflection will occur. Instead of computing the resulting load voltage and the magnitude of the reflection as done in the previous equations, we recognize that the voltage at the load depends only upon the previous voltage and the reflected signal arriving at the load. This process is called a *recursion* and we seek to write a formula for the load voltage when the *n*th reflection arrives. The voltage at the load will be the sum of the last voltage at the load and the voltage caused by the reflected signal from source. Introducing a time index, this is

$$V_L(2) = V_L(1) + V_{F2}(1 + K_L) \tag{9.13}$$

Noting that $V_{F2} = V_F K_L K_S$, where V_F was introduced in Equation 9.6, Equation 9.13 becomes

$$V_L(2) = V_L(1) + V_F K_L K_S \left(1 + K_L\right) \tag{9.14}$$

We then observe that each time a reflection arrives at the load its magnitude is equal to the previous reflection multiplied by $K_L K_S$. Therefore Equation 9.14 can be used to compute the load voltage at index n as

$$V_L(n) = V_L(n-1) + V_F (K_L K_S)^n \left(1 + K_L\right) \tag{9.15}$$

Finally, we recognize this as a geometric series[5] and rewrite it as

$$V_L(n) = V_F \left(1 + K_L\right) \frac{\left(1 - \left(K_S K_L\right)^n\right)}{\left(1 - K_S K_L\right)} \tag{9.16}$$

And the value of the first undershoot at $n = 2$ is

$$V_F \left(1 + K_L\right) \frac{\left(1 - \left(K_S K_L\right)^n\right)}{\left(1 - K_S K_L\right)} = 3.05\,(1+1)\frac{\left(1 - (-0.846 \times 1)^2\right)}{(1 - (-0.846 \times 1))} = 0.937\ \text{V} \tag{9.17}$$

Figure 9.7 shows the voltages at the oscillator output and the clock input.

Suppose the datasheet for the latch shows that $V_{IN}\text{Low}_{MAX} = 1.5$ V. This means that if a reflection causes the clock input to fall below 1.5 V, the latch would interpret this as a logic low. When the clock returns high, the latch would be erroneously clocked or *double-clocked*. Equation 9.17 and Figure 9.7 show that at 50 ns the latch would be clocked a second time.

One way to prevent double-clocking from reflections is to add a 120 Ω resistive termination at the clock input of the latch as shown in Figure 9.8a. Since the terminating resistor is equal to the characteristic impedance of the transmission line, there is no reflection at the load. If this technique is used, the oscillator must have the ability to drive the terminating resistor.

Another way to prevent double-clocking is to add a 110 Ω series resistance at the oscillator output so the series combination of the oscillator's source impedance and the 110 Ω equals the characteristic impedance of the

[5]This looks like mathematical trickery, but it's actually straightforward. If you write out a few terms of Equation 9.15 you'll see the geometric series. Then use a table of geometric series summations [2] for the formula for the sum.

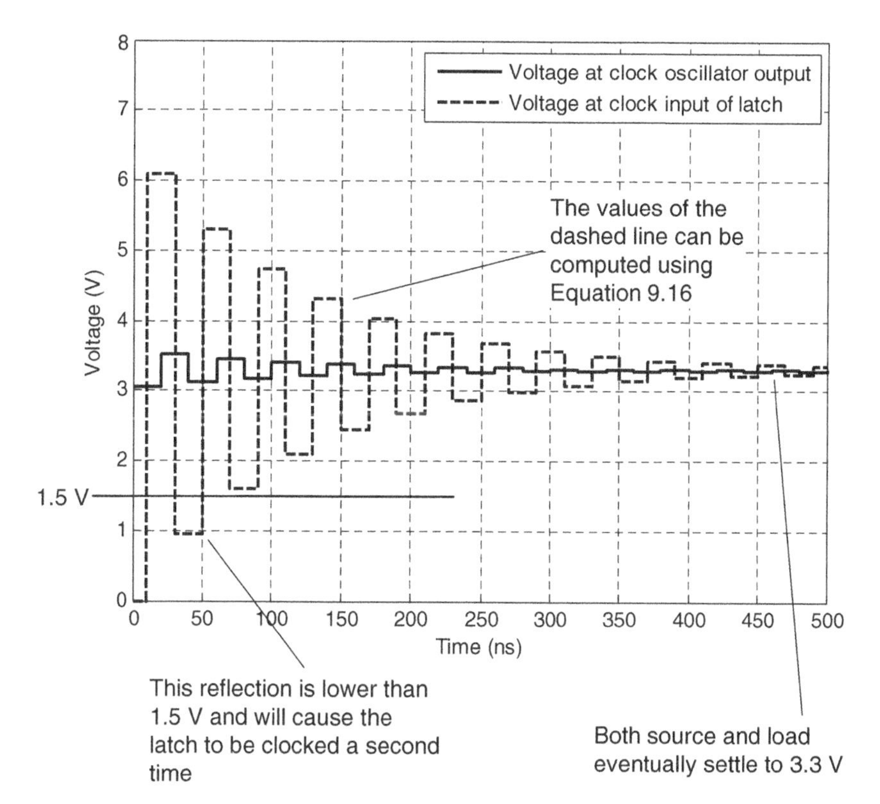

Figure 9.7. Waveforms for the source and load for the circuit shown in Figure 9.6. The latch is erroneously clocked by the first reflection because it pulls the clock input below 1.5 V.

transmission line. This results in a reflection at the load, but the reflection is eliminated when it returns to the source. This is shown in Figure 9.8b.

There are practical reasons for keeping the source resistance lower than the characteristic impedance of the transmission line [1]. Under these conditions, reflections will still occur, but they can be reduced so that double-clocking doesn't occur. Figure 9.8c shows 20 Ω in series with the oscillator output. Since the reflection no longer pulls the clock input below 1.5 V, the latch works as expected.

Undesirable effects of signal reflections are associated with the *rise and fall times* of the driving circuit, not the *frequency of the signal*. The key to identifying potential reflection problems is to check whether the rise and fall times are comparable to the two-way propagation delay[6] of the printed circuit

[6] Section 9.4 discusses propagation delay.

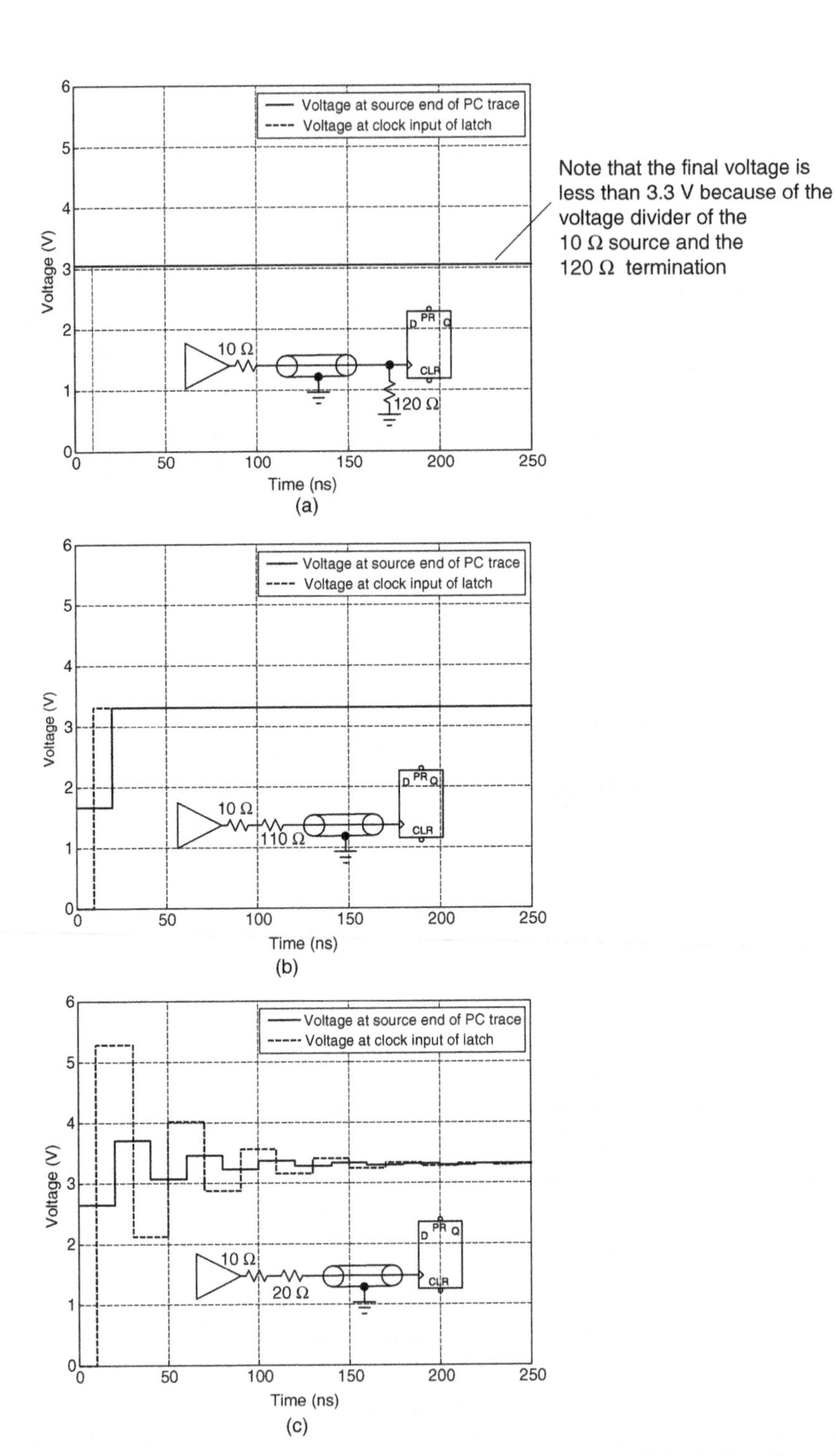

Figure 9.8. Waveforms for the oscillator output and clock input of the latch in Figure 9.6. In (a), a 120 Ω terminating resistor is placed at the clock input. In (b), a 110 Ω terminating resistor is placed in series with the 10 Ω oscillator output resistance. In (c), a 20 Ω resistor is placed in series with the oscillator output.

trace. If this is the case, reflections should be considered in the circuit design and printed circuit board layout.

9.4 EFFECT OF SIGNAL REFLECTIONS ON NARROWBAND SIGNALS

The concepts and equations developed in Section 9.2 are valid for any signal. Section 9.3 discussed the behavior of digital signals. For the remainder of this chapter we consider *narrowband signals*. An example of a narrowband signal is a signal from the radio in a police car which is centered around 860 MHz but has a bandwidth of only 25 kHz. For our purposes, we treat narrowband signals as simple sinusoids.

The reflection coefficient at the load is given by Equation 9.5. The reflection coefficient looking into a lossless cable from the source is

$$K_{\mathrm{L\ Source}} = \mathrm{e}^{-j2\beta d} K_{\mathrm{L\ Load}} \tag{9.18}$$

where β is the phase constant defined below and d is the length of the cable in meters. Equation 9.18 shows that the effect of the lossless cable is a periodic rotation of the reflection coefficient.[7]

β is defined as

$$\beta = \frac{2\pi f}{v} \ \mathrm{rad/m} \tag{9.19}$$

where f is the frequency of the signal and v is the velocity of signal propagation through the cable. The typical propagation velocity for many plastic dielectrics is 0.66 times the speed of light or $0.66 \times 3 \times 10^8$ m/s. This factor will be used for all examples in this chapter.

Example 9.1. Determine the impedance and reflection coefficient looking from the source into the cable for the circuit of Figure 9.9. The frequency is 67 MHz.

Solution. We solve this problem using the equations above and Matlab as our calculator as shown in Figure 9.10.

The computed input impedance is $64.6 + j75.5\ \Omega$ and the reflection coefficient is $0.392 + j0.401$.

[7]This suggests that if reflection coefficient was plotted on an x–y plot, the effect of inserting a cable would be a circular rotation. Could this have been Phillip Smith's motivation when he invented the Smith chart in 1939?

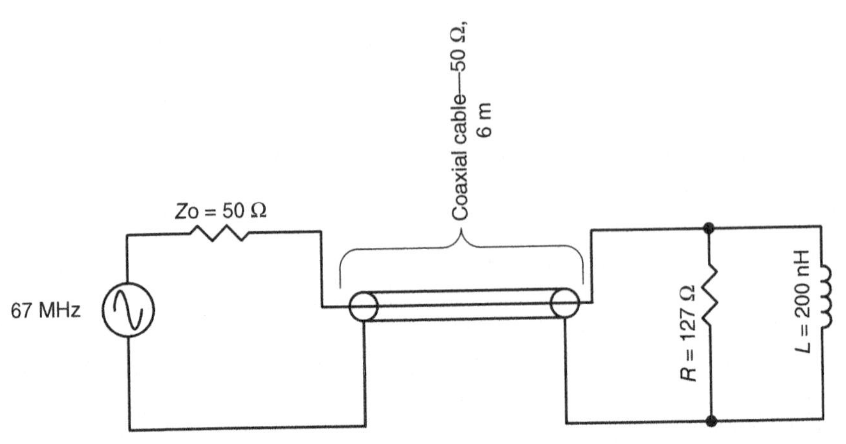

Figure 9.9. Circuit for Example 9.1. The impedance seen by the source is different than the impedance of the load.

Practical coaxial cables typically have a small amount of loss, meaning that some of the forward and reverse energy is dissipated as heat thereby attenuating the signals. Cable loss is specified in decibel per meter.[8] We represent the cable loss as a single attenuator as shown in Figure 9.11.

We write Equation 9.20 by starting at the source and incorporating all the effects that comprise the reflected wave:

$$K_{\mathrm{L(source)}} = \frac{V_{\mathrm{R}}}{V_{\mathrm{F}}} = \frac{V_{\mathrm{F}}\mathrm{e}^{-j\beta d_1} A\mathrm{e}^{-j\beta d_2} K_{\mathrm{L}} \mathrm{e}^{-j\beta d_2} A\mathrm{e}^{-j\beta d_1}}{V_{\mathrm{F}}} = A^2 K_{\mathrm{L}} \mathrm{e}^{-j2\beta(d_1+d_2)} \quad (9.20)$$

This equation shows that the effect of attenuation is to reduce the magnitude of the reflection coefficient while leaving the phase unchanged. The equation

```
1   F = 67e6;     % Frequency
2   w = 2*pi*F;
3   Zo = 50;                    % Cable impedance
4   R = 127;                    % Load resistance
5   L = 200e-9;                 % Load inductance
6   XInd = 1j*w*L;              % Inductor reactance
7   ZL = R * XInd/(R+XInd); % Total load impedance is the parallel combination.
8   Kl = (ZL-Zo)/(ZL+Zo);    % Reflection coefficient at load.
9   d = 6;                      % Cable length in meters.
10  B = 2*pi*F/(0.66*3e8);   % Phase constant
11  KIn = exp(-1j*2*B*d)*Kl;    % Reflection coefficient looking into cable from source.
12
13  ZIn = Zo * (1+KIn)/(1-KIn); % Impedance looking into cable from source.
```

Figure 9.10. Matlab code for Example 9.1.

[8] A common RF cable, RG-58, has loss of about 0.11 dB/m at 50 MHz.

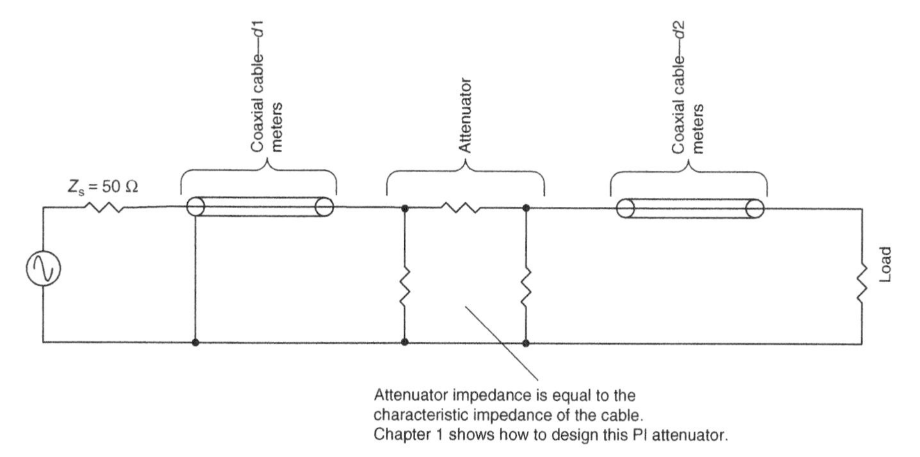

Figure 9.11. Cable and load with attenuation. The attenuator can be a single device as shown or it can be distributed through the cable. The placement of the attenuator does not affect the impedance seen by the source.

also shows that the attenuator can be placed anywhere on the transmission line without affecting the reflection coefficient seen by the source.

A final useful formula is the relationship between the reflection coefficient and the ratio of forward power, P_F, to reverse power, P_R:

$$P_F = \frac{|V_F|^2}{Z_O}, \quad P_R = \frac{|V_R|^2}{Z_O} \tag{9.21}$$

$$\frac{P_R}{P_F} = \frac{|V_R|^2}{|V_F|^2} = |K_L|^2 \tag{9.22}$$

Example 9.2. A power amplifier with 50 Ω output impedance drives a load of 85 + *j*32 Ω through a 50 Ω coaxial cable. The forward power is 250 W. Determine the reverse power.

Solution. Convert the load impedance to a reflection coefficient using Equation 9.5:

$$K_L = \frac{(Z_L - Z_O)}{(Z_L + Z_O)} = \frac{85 + j32 - 50}{85 + j32 + 50} = 0.299 + j0.166 \tag{9.23}$$

Use Equation 9.22 to compute the reverse power

$$P_R = |K_L|^2 P_F = |0.299 + j0.166|^2 \times 250 = 29.2 \text{ W} \qquad (9.24)$$

Note that we did not need to know the cable length for this problem. This is because the effect of the cable is to rotate the reflection coefficient around a circle. The rotation does not change its magnitude.

9.5 THE SMITH CHART

The Smith chart shown in Figure 9.12 was invented as a computational aid for working with cables and loads over 70 years ago. Today we have computers and calculators to perform the computational tasks, but the Smith

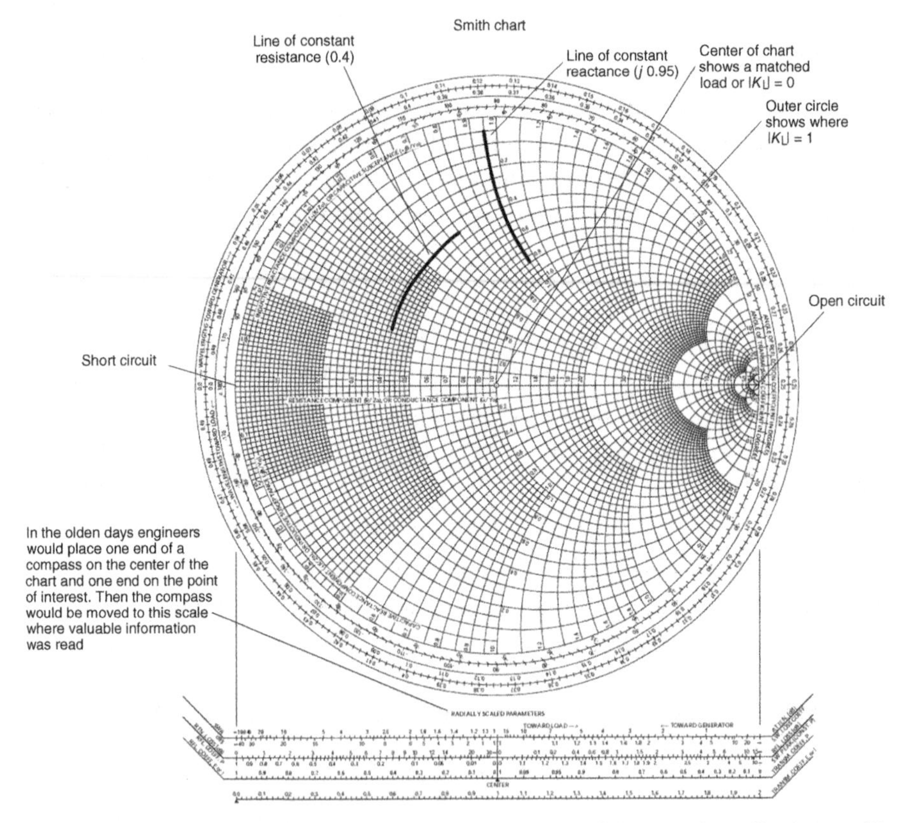

Figure 9.12. The Smith chart is fundamentally an *x–y* plot of the complex reflection coefficient K_L. It provides valuable insight when working with transmission lines and mismatched loads.

chart remains a valuable tool because it provides intuitive insight into the material discussed in Section 9.4. At first the Smith chart looks formidable, but in reality it's easy to use. Blank Smith charts and excellent Smith chart calculators can be readily downloaded from the Internet.

The basis of the Smith chart is that it is an x–y plot of the complex reflection coefficient K_L. The numbers on the interior of the Smith chart represent normalized resistances and reactances where the normalization constant is the impedance of the system (usually 50 or 75 Ω). For example suppose $K_L = 0.3 + j0.5$. We plot the reflection coefficient in Figure 9.13 using rectangular

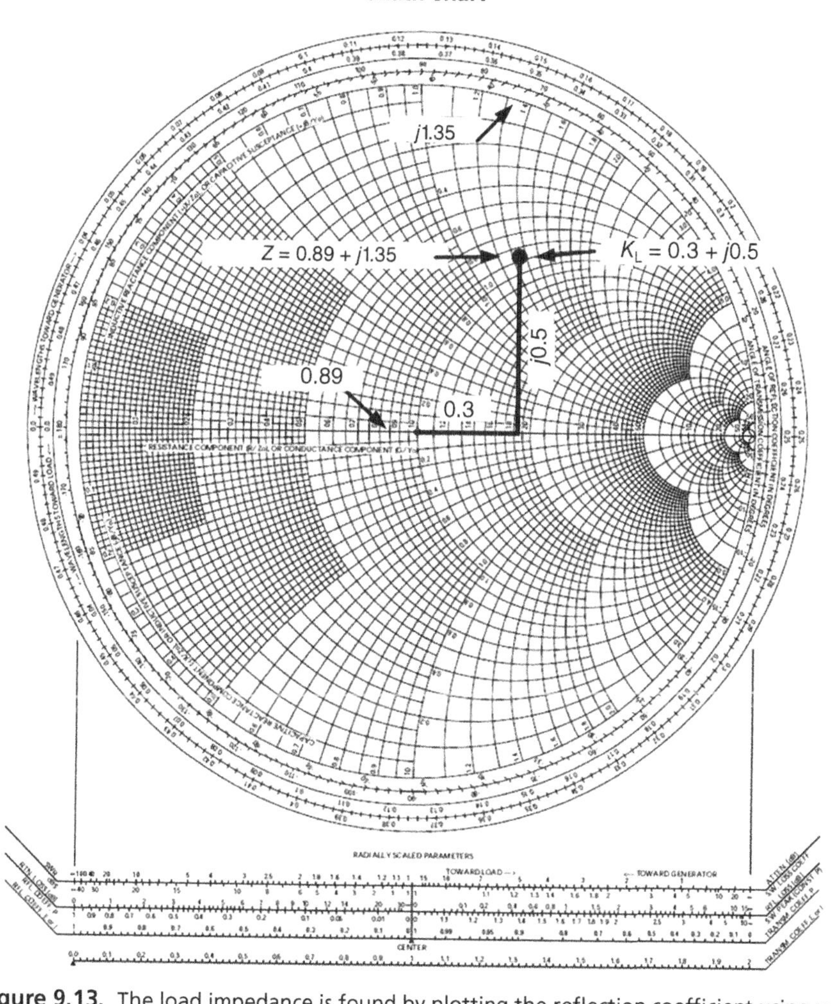

Figure 9.13. The load impedance is found by plotting the reflection coefficient using x–y coordinates and then using the contours of the chart to read the impedance.

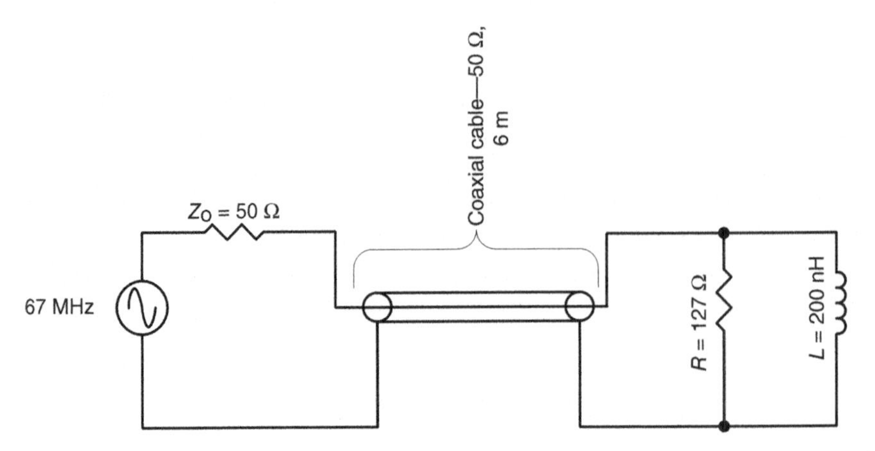

Figure 9.14. Circuit for Example 9.3. The impedance seen by the source can be computed as in Example 9.1 or determined graphically using the Smith chart.

coordinates and read the normalized impedance from the numbers on the chart and then verify our result using Equation 9.4:

$$Z_L = Z_O \frac{(1 + K_L)}{(1 - K_L)} = Z_O \frac{(1 + (0.3 + j0.5))}{(1 - (0.3 + j0.5))} = Z_O (0.892 + j1.35) \quad (9.25)$$

Example 9.3. Find the impedance seen by the source for the circuit of Example 9.1 using a Smith chart.

Solution. The circuit is shown in Figure 9.14.

The lines of Matlab shown in Figure 9.15 compute the normalized load impedance $Z_{Ln} = 0.776 + j1.17\ \Omega$ and we plot this point as A in Figure 9.16. We then use Equation 9.19 to compute the phase constant:

$$\beta = \frac{2\pi F}{v} = \frac{2\pi \times 67 \times 10^6}{0.66 \times 3 \times 10^8} = 2.1261 \text{ rad/m} \quad (9.26)$$

```
1   F = 67e6;              % Frequency
2   w = 2*pi*F;
3   R = 127;               % Load resistance
4   L = 200e-9;            % Load inductance
5   XInd = 1j*w*L;         % Inductor reactance
6   ZL = R * XInd/(R+XInd); % Total load impedance is the parallel combination.
7   ZLn = ZL/50;           % Normalize load impedance for plotting on Smith chart.
```

Figure 9.15. Matlab code for Example 9.3.

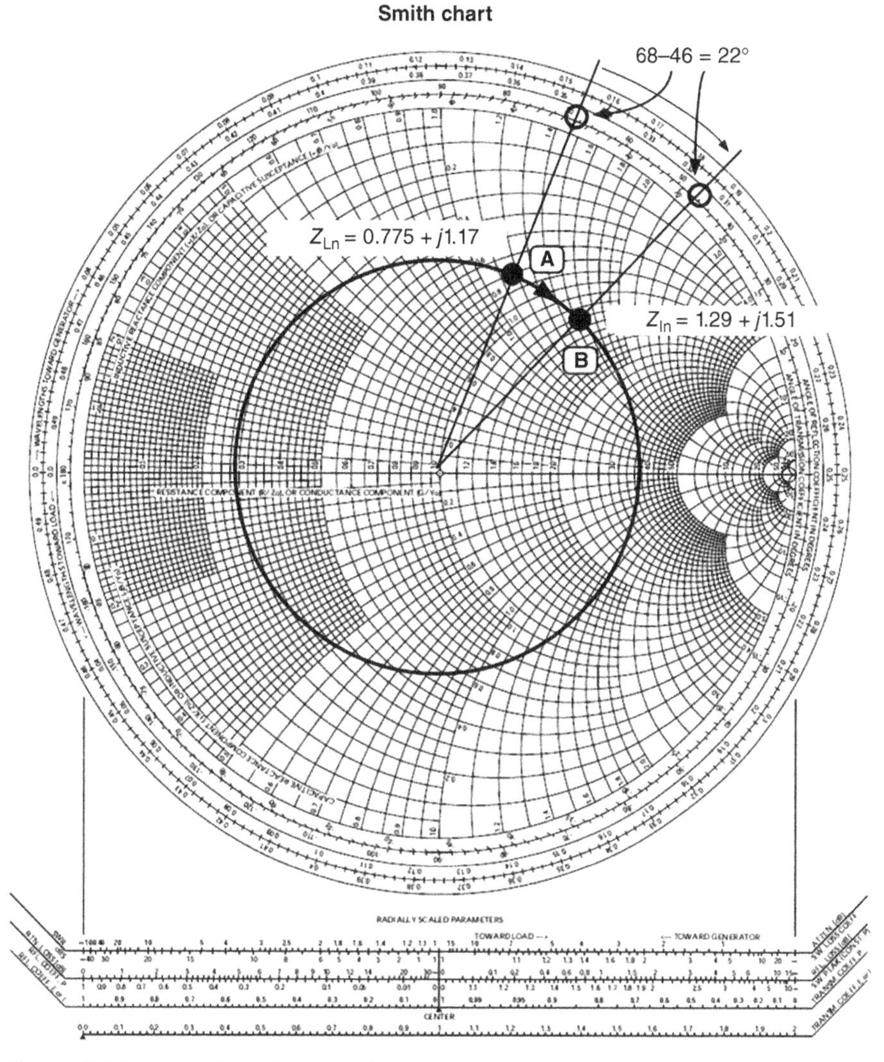

Figure 9.16. Smith chart for Example 9.3. Point A is the load. The effect of the cable is four clockwise rotations plus an additional 22°.

Convert the rotation angle for the reflection coefficient:

$$-2\beta d = -2 \times 2.1261 \times 6 \times \frac{360}{2\pi} = -1462° \tag{9.27}$$

This corresponds to four full clockwise revolutions around the chart followed by an additional 22°. The scale labeled "angle of reflection coefficient" on the outer ring of the chart simplifies rotation since it is labeled in degrees.

The normalized impedance seen by the source is point B. The normalized value from the chart is $1.29 + j1.51\ \Omega$. To denormalize multiply by 50 to get $64.5 + j75.5\ \Omega$.

So far we have plotted only normalized impedance on the Smith chart, but the chart can be used to plot normalized admittance also. In other words the same chart can be used as an impedance chart or an admittance chart. This is useful for problem solving because if the chart is an impedance chart, then series impedance can be added easily. If the chart is an admittance chart, then parallel components can be added easily. When solving problems using the Smith chart it's often beneficial to switch between using the chart for impedance or admittance. Say we have plotted an impedance point on an impedance chart and we would like to use the chart for admittance. This is easily done by simply rotating the point by 180°.[9] This is shown in the following example where we use the Smith chart to solve a practical problem.

Example 9.4. Given a load of $10 - j70\ \Omega$ at 2 MHz, add components so that the combination of the added components and the load is 50 Ω. The added components may not dissipate power.

Solution. This is called a *matching*[10] problem because the added components will match the load to a 50 Ω source and cable. When the load is matched then all of the power generated by the source is dissipated in the resistive part of the load and there are no reflections.[11]

The normalized load value is

$$\frac{10 - j70}{50} = 0.2 - j1.4\ \Omega \tag{9.28}$$

The Smith chart of Figure 9.17 allows us to visualize our strategy. The matching network may not consume power so the choice of components is restricted to inductors and capacitors. Point A on the Smith chart shows the load. Starting at A we add series inductance by following the $R = 0.2$ contour until we get to point B which when rotated 180° lies on the $R = 1$ circle. We then convert to admittance by rotating B 180° taking us to point C. From C we see that if capacitive susceptance[12] is added, we will follow the $R = 1$

[9]You can easily convince yourself of this by computing the admittance as the reciprocal of the impedance and plotting it on the chart.

[10]Impedance matching calculators such as [3] are available online.

[11]This is especially important when driving an antenna with a RF power amplifier. Not only do we want the antenna to radiate as much power as possible, we also don't want to dissipate the reflected power in the power amplifier.

[12]Susceptance is the imaginary part of admittance.

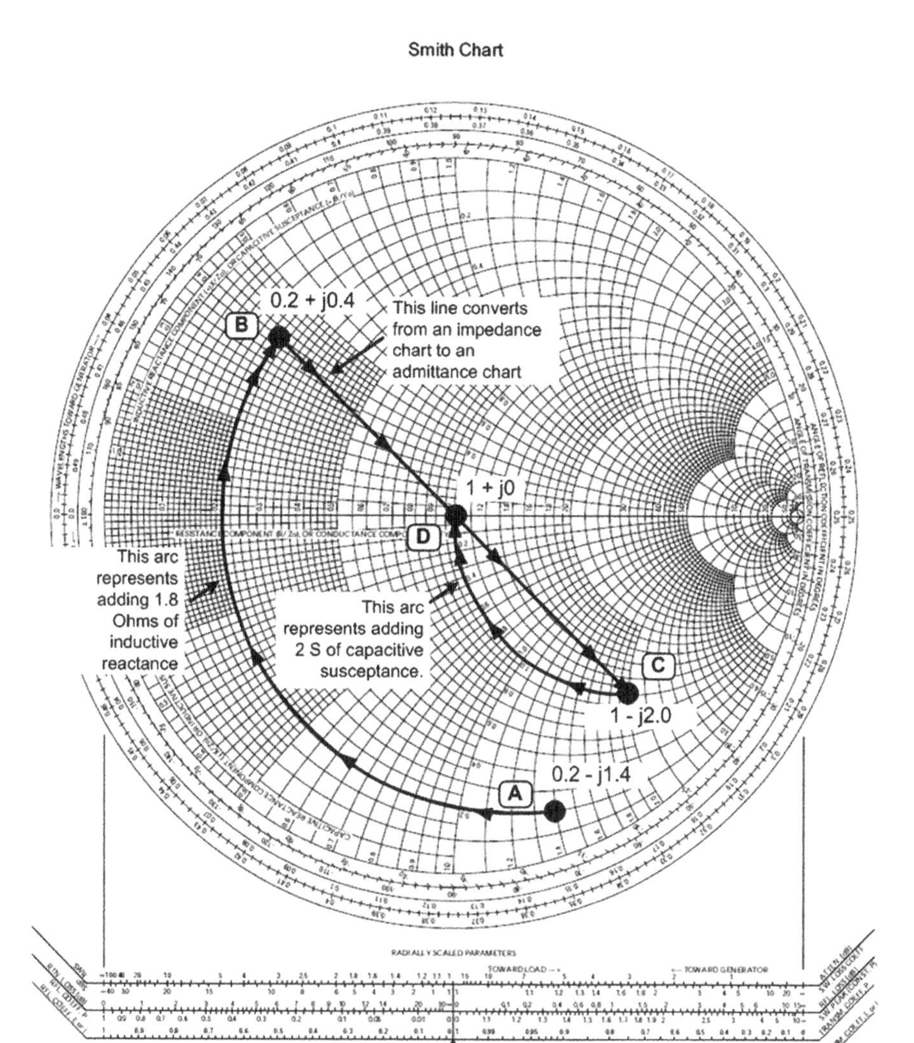

Figure 9.17. Smith chart for Example 9.4. Points A and B are impedances. Point C is an admittance. Point D represents $1 + j0\Omega$ so it can be considered impedance or admittance.

contour to D which is the desired goal of $1 + j0\ \Omega$. With this strategy in place we work the problem.

To get from A to B, add 1.8 Ω of normalized inductive reactance in series. At 2 MHz this is

$$L = \frac{50 \times 1.8}{2\pi \times 2 \times 10^6} = 7.16\ \mu\text{H} \tag{9.29}$$

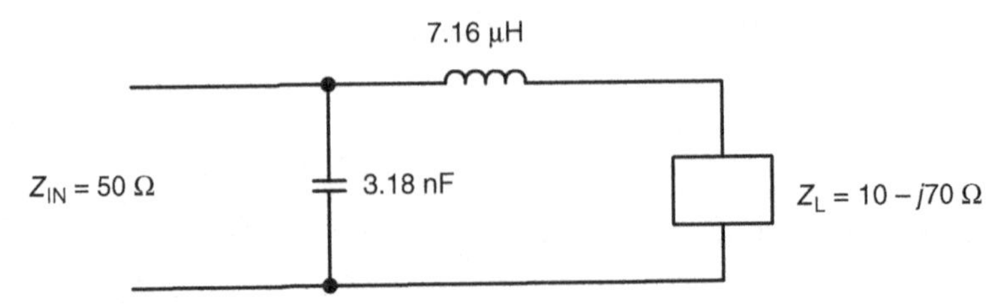

Figure 9.18. Matching network and load for Example 9.4. The result of adding the inductor and capacitor is that reflections on the transmission line are eliminated and all power from the source is delivered to the load.

Since the next step is to add admittance, point B is rotated 180° resulting in point C.

To get from C to D add two units of normalized capacitive susceptance to ground. The normalized capacitive reactance is the inverse of the susceptance:

$$X_C = -\frac{50}{2} = -25\ \Omega \tag{9.30}$$

And the capacitance is

$$C = \frac{1}{2\pi \times 2 \times 10^6 \times 25} = 3.18\ \text{nF} \tag{9.31}$$

The matching network and load are shown in Figure 9.18. We check our work by computing the input impedance at the input terminals. The series reactance of the load and inductor is

$$10 - j70 + j2\pi \times 2 \times 10^6 \times 7.16 \times 10^{-6} = 10 + j19.975\ \Omega \tag{9.32}$$

The reactance of the capacitor is

$$\frac{1}{j2\pi \times 2 \times 10^6 \times 3.18 \times 10^{-9}} = -j25.02\ \Omega \tag{9.33}$$

Combining the branches in parallel gives

$$Z_{IN} = \frac{(10 + j19.975)(-j25.02)}{10 + j19.975 - j25.02} = 50\ \Omega \tag{9.34}$$

which is the desired value.

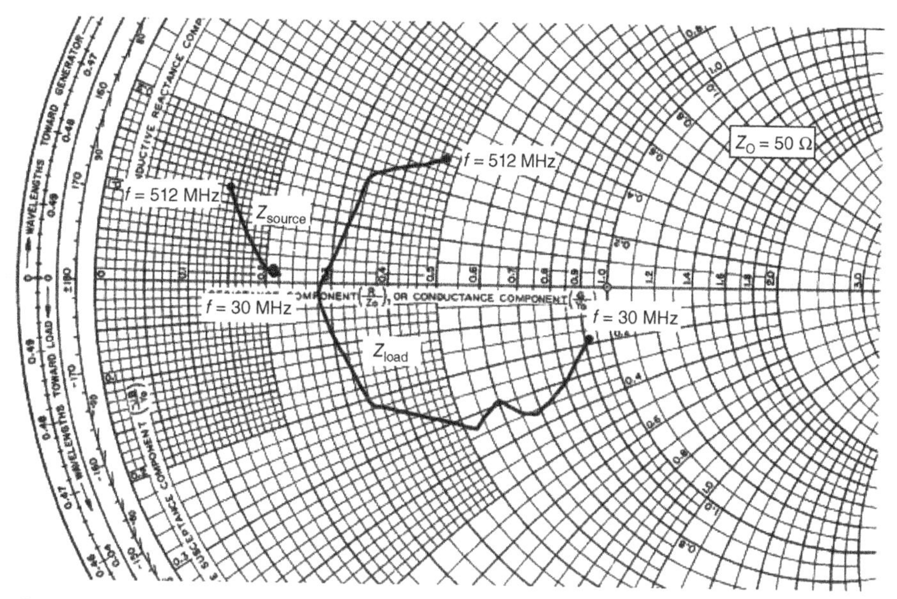

Figure 9.19. Smith chart displaying device impedance versus frequency. (Copyright 2013 Freescale Inc., used with permission)

9.6 USING THE SMITH CHART TO DISPLAY IMPEDANCE VERSUS FREQUENCY

The Smith chart can also be used for displaying impedance as a function of frequency. In Example 9.4 we noted that the center of the Smith chart represents a perfect match. This is no surprise. Equation 9.22 shows that the ratio of reflected to forward power is simply the magnitude of the reflection coefficient squared. In other words the radial distance from the center of the Smith chart is an indication of how well an impedance is matched to the reference impedance. Manufacturers often display the impedance of their devices versus frequency on a Smith chart as shown in Figure 9.19 so designers can visually see how well the device is matched at their frequency of interest.[13]

9.7 FINAL COMMENTS REGARDING THE SMITH CHART

If you can remember a few things about the Smith chart, then you'll be prepared to use it with only a quick review.

[13]Note that when working problems on the Smith chart a single frequency is used. When the chart is used to display data, the plot represents device behavior at different frequencies.

1. There is a one-to-one mapping between reflection coefficient and impedance.
2. The Smith chart is an x–y mapping of the complex reflection coefficient.
3. The quality of a match is represented by the radial distance from any point to the center of the chart.
4. The effect of a cable is a rotation around the chart.

PROBLEMS

9.1 A digital gate is connected to the clock input of a latch as shown in Figure 9.20. The rise time for the gate is 1 ns and the one-way propagation delay is 30 ns. The printed circuit trace has a characteristic impedance of 88 Ω. The output impedance of the driver is 15 Ω and there is a 1 kΩ termination at the clock input. At time $t = 0$ the driver switches from 0 to 3.0 V.

(a) Is this a situation where reflections should be considered? Why?
(b) What is the source voltage immediately after the driver switches?
(c) What is the load reflection coefficient?
(d) What is the voltage at the load immediately after the first reflection?
(e) What is the voltage of the reflected signal?
(f) What are the voltages at the gate output and latch input after all reflections have dissipated?

9.2 A digital gate is connected to the clock input of a latch as shown in Figure 9.21. The rise time for the gate is 1 ns and the one-way propagation

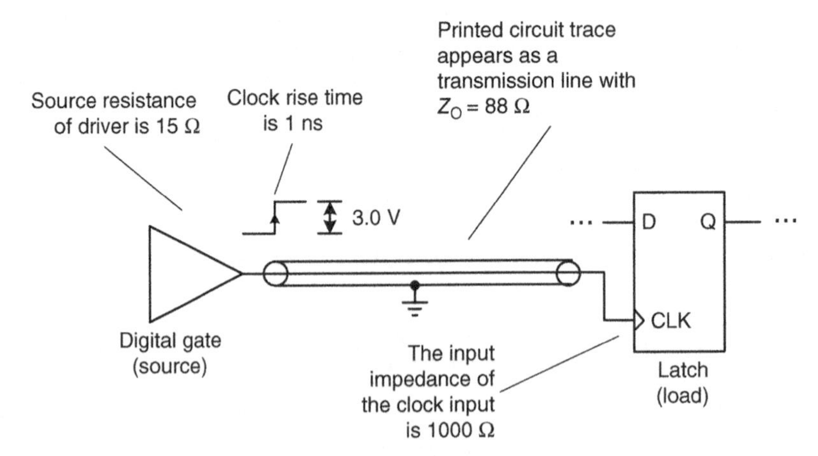

Figure 9.20. Circuit for Problem 9.1.

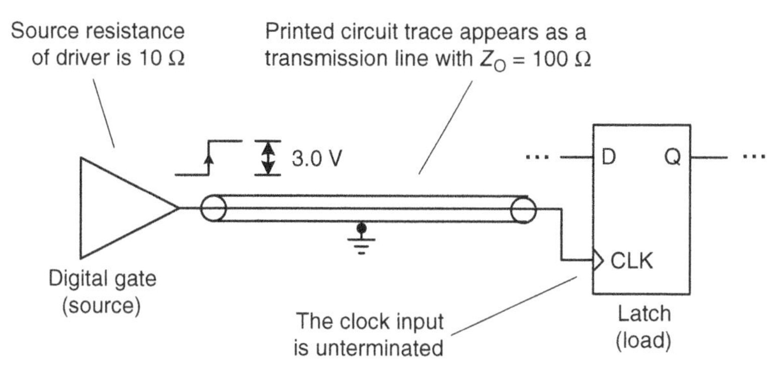

Figure 9.21. Circuit for Problem 9.2.

delay is 25 ns. The printed circuit trace has a characteristic impedance of 100 Ω. The output impedance of the driver is 10 Ω and the clock input is unterminated. $V_{IN}Low_{MAX}$ for the latch is 2.0 V. At time $t = 0$ the driver switches from 0 to 3.0 V. *Hint:* Create an Excel worksheet based on Equation 9.16. Use the "goal seek" function for parts (c) and (d).

(a) Will undershoot cause double-clocking of the latch?

(b) How much time must elapse before the output is guaranteed to be greater than 2.9V?

(c) What is the minimum series source termination that will prevent double-clocking?

(d) What is the maximum shunt load termination that will prevent double-clocking?

9.3 Determine the impedance, Z_{IN}, and reflection coefficient for the circuit shown in Figure 9.22. The frequency is 41 MHz. The characteristic impedance is 50 Ω.

9.4 A 5600 pF capacitor is connected to a 2 m, 50 Ω coaxial cable as shown in Figure 9.23. The frequency is 150 kHz. Use a Smith chart to show

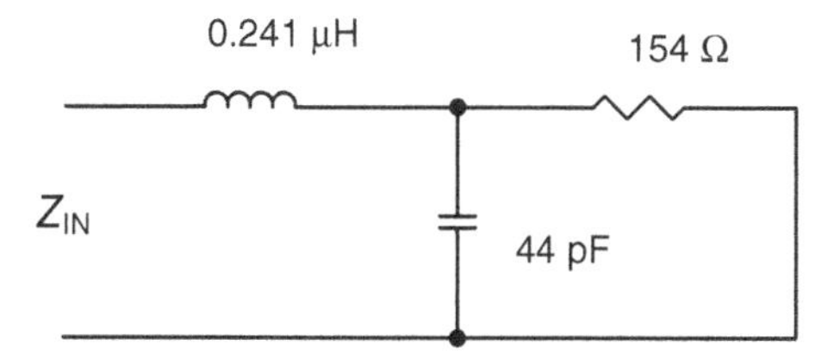

Figure 9.22. Circuit for Problem 9.3.

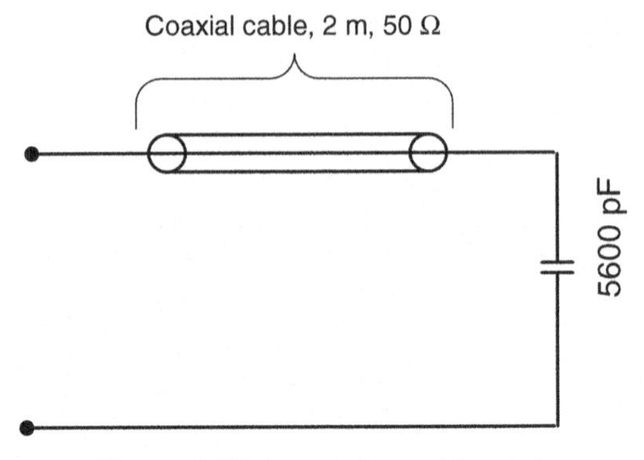

Figure 9.23. Circuit for Problem 9.4.

the impedance looking into the cable. Comment whether a Smith chart was needed for this problem and why.

9.5 The cable in the circuit of Figure 9.24 has loss as shown. Determine the impedance looking into the cable from the source. Use a Smith chart

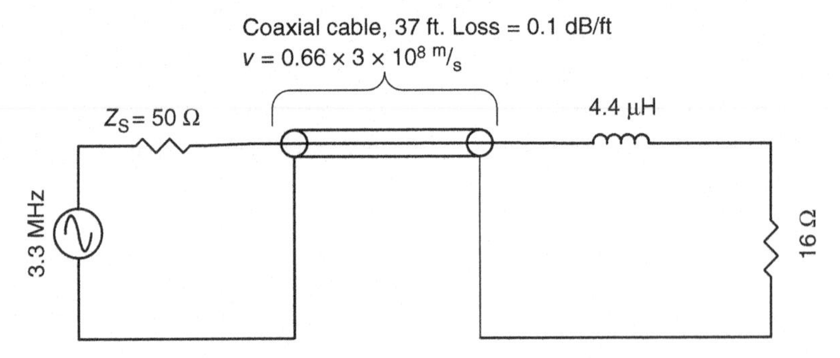

Figure 9.24. Circuit for Problem 9.5.

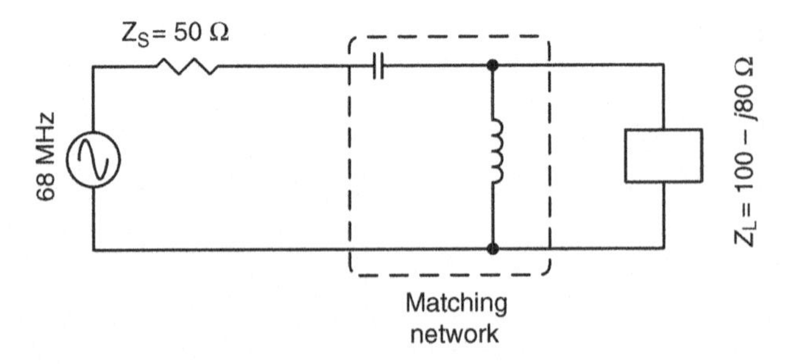

Figure 9.25. Circuit for Problem 9.6.

and check your work by calculating the result. *Hint:* Either use Equation 9.20 or use the attenuation scale at the lower-right corner of the Smith chart.

9.6 Use a Smith chart to determine matching components for the circuit shown in Figure 9.25 using the topology shown.

REFERENCES

1. H. W. Johnson, *High-Speed Digital Design*, Prentice Hall, 1993 (Accessed 11-29-2013.)
2. Wikipedia. *Geometric Series*. Available: http://en.wikipedia.org/wiki/Geometric_series (Accessed 11-29-2013.)
3. J. Wetherell. *Impedance Matching Network Designer*. Available: http://home.sandiego.edu/~ekim/e194rfs01/jwmatcher/matcher2.html (Accessed 11-29-2013.)

10

GETTING A JOB—KEEPING A JOB—ENJOYING YOUR WORK

A satisfying, rewarding and meaningful career requires both technical and interpersonal skills. The previous nine chapters focused on the technical skills you'll need to succeed in your job interviews and become an immediate contributor in the workplace. This chapter focuses on interpersonal skills and behaviors that will help you get a job and then enjoy a successful career.

What is a successful electrical engineer? The author's friend, mentor, and colleague, Dave, is the most successful engineer I know. When project teams are formed, Dave is in high demand because of his excellent teamwork skills. When there is a technical problem, Dave is consulted because of his broad knowledge and excellent troubleshooting skills. When engineers are just stumped, they come to Dave because of his ability to think "outside the box." When engineers need career advice, they come to Dave because he is an excellent listener and hands out advice wisely and respectfully. When Dave needs help, he is equally comfortable learning from his colleagues, reading technical material from manufacturers, or opening up the latest text-book. Dave is a great source of stories about climbing mountains, starting

Ten Essential Skills for Electrical Engineers, First Edition. Barry L. Dorr.

companies, climbing 2000-ft antenna towers, running from bears, shanking golf balls, and participating in many engineering development efforts. Dave has enjoyed his career and contributed positively to the careers of many others since he graduated from college in 1968. Dave is 74 years old and, despite the requests to play more golf with his friends, continues to be an active contributor to our department.

This chapter contains advice from the author, from Dave, and from successful engineers the author is fortunate to call friends; some are highly technical, some manage people and projects, and some market products. This chapter is a collection of suggestions intended to supplement the traditional career advice that can readily be found on the Internet or elsewhere. These items describe qualities and behaviors seen in the *best* interview candidates and the *best* engineers. This information will help you positively distinguish yourself from other job applicants and then distinguish yourself in the workplace. Not everything you read in this chapter will be right for you. Pick each item up, examine it, and determine if it will help you be your best.

Companies hire *engineers* because they need engineering talent to solve technical problems. Companies hire *people* to effectively deliver those skills to the workplace. The first section of this chapter focuses on how you can effectively market yourself to companies as a candidate with strong technical and interpersonal skills. Demonstrating these skills during the hiring process is essential to getting your first job.

The second section of the chapter shows how to quickly develop skills that will make you a valuable part of any engineering department. This will allow you drive your career in the direction you want it to go and also minimize the chances you'll be a target when the company lays off engineers. These are simple suggestions and might seem obvious, but they can be the difference between an excellent engineer and a mediocre one.

The third section of the chapter acknowledges that the nearly 100,000 hours of our life spent working should be satisfying, rewarding, and meaningful. As a recent graduate, this material won't be immediately useful, but as your career progresses it will provide insights on how to keep your work experience fresh and exciting while keeping your skills in demand.

10.1 GETTING A JOB

When a company has a job opening, the process in Figure 10.1 is used to narrow the field of applicants. The large number of resumes received by a company is shown at the top of the figure. Some of these resumes will be sent by individuals, but many of them will be sent automatically by computers at staffing firms that use software to align phrases and terms on engineers'

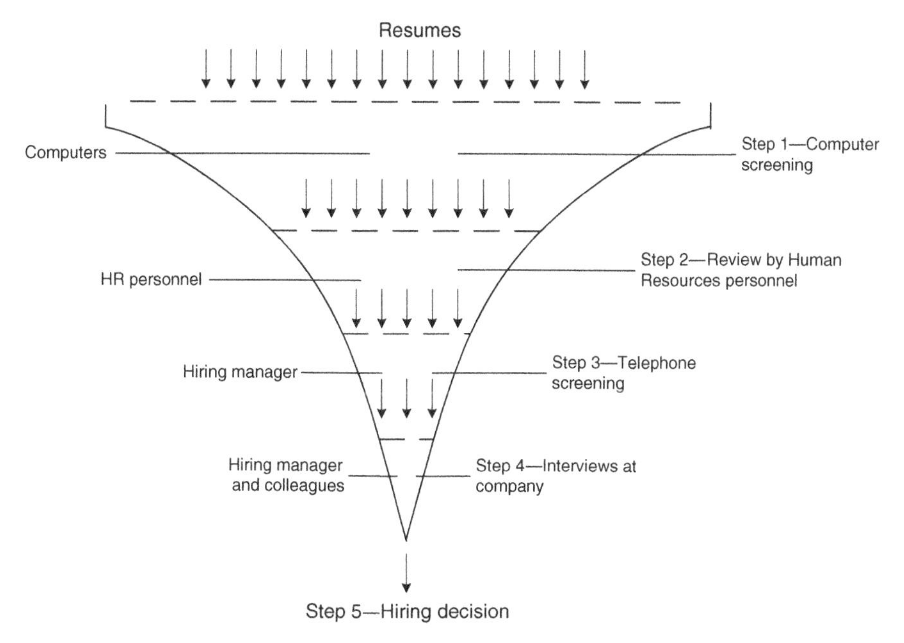

Figure 10.1. The corporate hiring process. This chapter gives suggestions for bypassing Step 1 because resumes of recent graduates rarely make it past this step.

resumes with those in the job posting. Due to the large volume of incoming resumes, the hiring company will often use its own computer software to screen the resumes as shown in Step 1. Some of the resumes will be written with this procedure in mind, and contain a large number of terms[1] that the engineer believes will get his resume through this process. Computer sorting results in a smaller collection of resumes to be reviewed by personnel in the company's Human Resources department. The Human Resources person typically has experience hiring engineers and will further screen the resumes based on specific input from the hiring manager. The result of this screening is about 30 resumes that will be passed to the hiring manager. The hiring manager will review the resumes and compare the experience described in the resumes to his needs. If the experience on a resume does not match the manager's needs, or if the hiring manager senses that it is overly rich with search terms but weak on experience, it will be rejected. The hiring manager will typically select about 10–15 candidates for interviews.

It is expensive for a company to conduct onsite interviews because of the time spent by the interviewers and possibly the candidate's travel expenses. Therefore the next step in the process is a telephone screen where the hiring

[1]Often taken directly from job postings.

manager interviews candidates for about 15–30 min. After the telephone screens, four or five candidates will be invited to the company for an onsite interview. Finally, the best candidate will receive a job offer.

As a recent graduate, the reason your resume will not make it through the company's computer screening process is that it will be in competition with those of experienced engineers. Companies recognize the value of new graduates and recruit them with on-campus interviews and job fairs. The best way to avoid the computer screening process is to take advantage of all the job placement assistance available at your school. Another way to avoid resume screening is to target companies carefully and send resumes to individuals within the company. This requires more homework on your part, but may result in a "dream job."

Students sometimes question why a company would hire a recent graduate instead of an experienced engineer—especially in an economy where many experienced engineers are competing for the same jobs. One reason against hiring recent graduates is that they don't provide immediate value, and the company simply doesn't have time to train them. This is a valid concern. The previous nine chapters have addressed it by reviewing practical skills that you can confidently demonstrate in an interview and then use as an immediate contributor in the workplace. If you've absorbed the skills in this book, your skill set will be competitive with many experienced engineers—especially those who have not maintained their skill sets.

There is an old joke about two guys who encounter a hungry cheetah in the African plains. One of them immediately starts putting on a pair of running shoes. The other asks "Why are you putting on those shoes? You can't outrun that cheetah." The first guy tightens the laces and replies "Of course not, I just have to outrun you." The point is that you don't need to be a superstar to get a good job. Instead, you need to positively differentiate yourself so that you are selected over the other applicants. The items in this chapter are suggestions for differentiating yourself in the interview process—and throughout your career.

10.1.1 Getting an Interview

The process of getting an interview begins well before you attend that first job fair or send out your first resume.

Take the Fundamentals of Engineering/Engineer In Training (FE/EIT) exam—Taking and passing this exam differentiates you from other candidates. It also shows an employer that you retained what you learned in your undergraduate courses. It also allows you to take the Professional Engineer (PE) exam within a year after college which will raise your professional collateral throughout your career.

Get some experience—If you worked as a co-op student or engineering intern you will be able to show some experience on your resume. If you don't have experience, consider doing a project on your own. If you aren't sure what to build, pick an example from this book that you find interesting and aligns with your career interests; then get it working in your school's senior project laboratory. Any time you make something work you gain experience that employers will value. Take some pictures, have them available on your phone and make sure to include this experience on your resume.

Attend job fairs—One of the best ways to bypass the computerized resume sorting process is to attend job fairs at your school. Companies frequently send hiring managers to college job fairs because they are specifically interested in recent graduates. These managers will often accept resumes and conduct interviews on the premises. If you interview well, you'll be invited to interview at their facility. When you go to the job fair, bring your resume, dress appropriately, and be prepared for an interview.

Send resumes to a carefully chosen set of companies—Sending a resume to every company with an available position will likely result in many computerized rejections as shown in Step 1 of Figure 10.1. It is much more effective to target companies where you have a particular interest and possibly some applicable experience.

Send your resume to the right person—You can sometimes bypass the computer resume sorting process by sending a resume and cover letter directly to a corporate executive, the hiring manager, or the Human Resources department. Corporate websites of small- and mid-size companies often provide names and e-mail addresses for members of the executive management team. If you send a well-written cover letter and resume to the director of engineering expressing your interest and asking for an interview, he may notice that you have some special quality or skill, she may believe that you would make a good addition to the department, or he may just recall that he was a recent graduate once and decide to help by placing you in contact with someone he knows. Don't send e-mail, because it's too easy for the recipient to simply delete it. A letter has a better chance of getting passed to the right person with hand-written or verbal comments.

Always accompany a resume with a cover letter—A cover letter shows that you have a particular interest in a company and differentiates you from applicants who simply send a resume. Your cover letter should show that you have researched the company, state why you're interested in working there, and include any experience that may be relevant to their business. If you're unsure of your spelling or grammar, have someone else check it for you. Don't send a "one-size-fits-all" cover letter. Personalize it for the job you're applying for.

Contact companies even if they are not hiring—Companies are always looking for good people. If you send a well-written cover letter and resume to the right person, he/she may reconsider their staffing needs and hire you. The process of hiring an engineer is expensive for a company and usually entails sifting through hundreds of resumes and interviewing a dozen or more candidates. That's a lot of work. If a good candidate "drops out of the sky," it saves them money.

10.1.2 Preparing for an Interview

Research a company before the interview—Good interviewers frequently begin telephone and onsite interviews with the question "So what do you know about our company?" The amount of research you've done tells them whether you are interested in the company or just in getting a job. Learn about some of their products and relate them to your interests and abilities.

Get your suit cleaned and pressed—Even though the technical interviewers may be wearing blue jeans and T-shirts, you as the candidate are expected to be well dressed. If you do well in the interview, you may be introduced to some of the executives in the organization who wear suits regularly and will appreciate that you are similarly attired.

Prepare to discuss anything on your resume—As an interviewer, it's disheartening to ask a student about something on his/her resume and get the response "I did that, but I don't remember much about it." This is often the case with team projects where the student includes the project on his/her resume but only did a small part of it. Doing a small part is perfectly acceptable, but you should be ready to describe what you did, what your team members did, how you worked together, and what you learned. For example, students frequently include "Matlab" on their resume, but can't remember the syntax of basic operations. But if they discuss where and why they used Matlab and what problems it solved, then they've demonstrated sufficient familiarity. As you prepare your resume, make sure you can say something meaningful about every detail of it.

Mentally prepare to be interviewed by a team—Team interviews are becoming more common, but seeing a group of people in the room surprises some candidates.

Be prepared for the interviewer to ask you to discuss a subject of your choice—This is your chance to relate an experience where your unique skills, knowledge, and personality resulted in success. You could discuss a lab, a team project, your senior project, or a project you did to gain experience. An ideal story would show how you analyzed a problem, came up with a solution, executed your plan, made something work, and learned something valuable.

Your story can be simple. For example, your "problem" could be that you had no experience, your solution could be that you did an individual project in the senior project lab (such as the circuit from Section 5.6), and what you learned was how to build and debug a servo system.

It is common for interviewers to ask the candidate if he/she has any questions at the end of the interview. If you have not had a chance to relate your story, ask if you can mention "An experience that will give you an idea of how I approach and solve problems."

10.1.3 The Interview

As shown in Figure 10.1, your next step to securing your job will be a telephone screen. If you do well in the telephone screen you will be invited to the company for an onsite interview.

After your onsite interview, each interviewer evaluates you by filling out a form provided by the Human Resources department. You will be graded in 10–15 areas including technical knowledge, enthusiasm, and whether the interviewer feels you can make an immediate contribution. At the bottom of the forms are checkboxes for "hire" and "don't hire," and under the boxes is a field for comments. Your goal is for the interviewers to check the "hire" box and then follow up with a comment such as "The candidate had researched our company and wants to work here because he/she likes our technology. He/she has solid technical skills and will be an immediate contributor to our department."

Treat the phone screen as an interview—You must succeed in your phone screen to get invited for an onsite interview. Make sure you are located in a quiet place with an Internet-connected computer. Since the audio quality of a land-based telephone is better than a cell phone, use one if possible. Expect technical questions, and have pencil, paper, and calculator with you. If you look something up on the Internet during the interview and the interviewer hears the key taps, you could be rejected. Use the Internet only if the interviewer asks you to go to a specific website.

Be prepared to work problems—To evaluate your technical skills, the interviewer needs to ask you to work out problems. This surprises some candidates and causes them to "freeze." You should bring a pencil, eraser, clipboard with paper, and calculator to any interview. Arrive early and work out several problems immediately before the interview so you are "in the zone."

Be confident and enthusiastic—The interview is your chance to show the interviewers that you are going to be a valuable asset to their department. You need to show that you just received an excellent education, you know how to apply it, and you have prepared specifically for this interview. The enthusiasm of recent graduates is valuable in engineering departments. If you feel you offer enthusiasm, let it show.

Don't worry about being nervous— Interviewers expect candidates to be nervous. Feel free to admit it to the interviewer if it helps you. If, at the beginning of the interview, they ask if you would like a glass of water, accept it so your throat doesn't get too dry during the interview. If you are nervous, don't starve your brain of oxygen by taking shallow breaths.

If you'll be interviewed by a team, try to sit at the head or the foot of the table—When interviewed by a team, it is awkward if you have to turn your head to address each person. If possible, try to sit at the head or foot of the table so you can address everyone in the room simultaneously.

When you are asked something you don't know, don't give up or panic—Interviewers intentionally present difficult questions to gauge how candidates approach unfamiliar problems. If the problem is difficult because it uses terminologies or technologies that are unfamiliar to you, ask for clarification. Draw a picture or diagram that shows what they're asking. Work with your diagram and use the fundamentals. Talk as you think. Let them stop you or prod you through.

If the interviewer is inexperienced, help him/her out—In today's economy the interviewer is often your potential boss, and he/she may not be an experienced interviewer. Figure 10.2 shows the "middle ground" where interviewing is effective. By reading this book, you have done your part to prepare for an interview in this area. If the interviewer attempts to conduct the interview in the area to the right of the middle ground, then you will be presented with terms and technologies that will be completely unfamiliar, and you won't get the opportunity to show what you can offer. If presented with questions on the right side of Figure 10.2, respond by asking questions that relate the questions to the middle ground.

Another mistake made by inexperienced interviewers is to spend too much time telling you about the problems and needs of the company without giving you an opportunity to present yourself. When the interviewer fills out the review form he/she will realize that they failed to get the required information and you won't get hired. If you detect this, salvage the interview by politely interrupting and letting the interviewer know how you can help with the problems he/she is describing.

When doing problems, check your work—When working problems during an interview you will make errors—and interviewers expect that. If you can recognize that your answer is wrong, you'll demonstrate that you have the skills to objectively review your work and find errors on your own. For example, when doing any analog or digital circuit problem, always apply asymptotic analysis to check the frequency response.

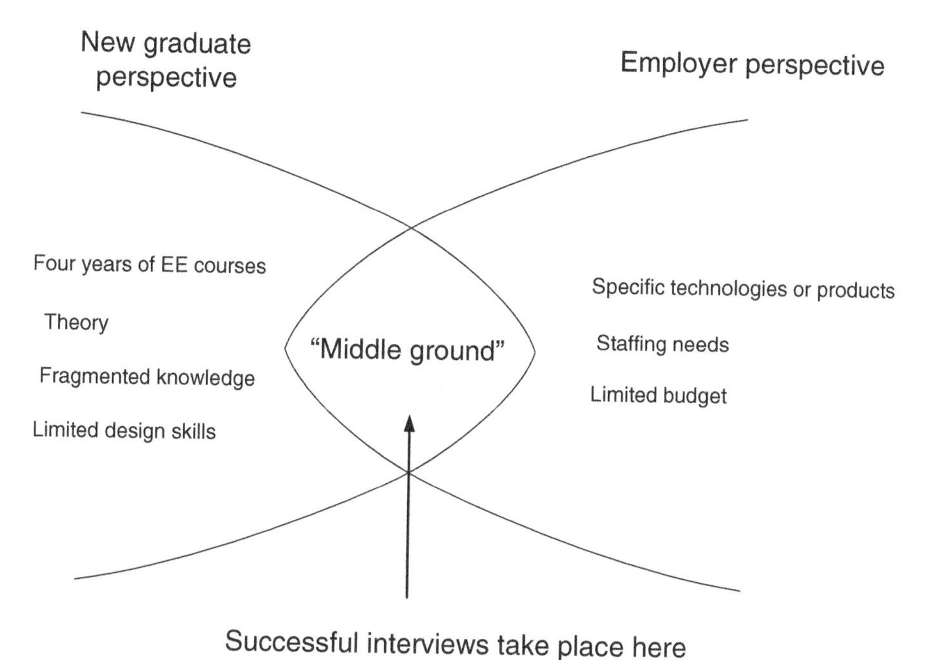

Figure 10.2. When interviews take place in the "middle ground," the company learns the most about the candidate and the candidate gets the best opportunity to show his/her skills.

When asked about team projects, make sure to credit your team members—The interviewer is probably more interested in the team dynamics than the specific result. Discuss your most successful project and show why you were a good team member.

Be prepared to ask the interviewer questions—Nearly all interviewers will conclude the interview by asking the candidate if he/she has questions about the company or the department. Asking the interviewer about his/her work will give you valuable insight into the department's tasks and how they do them. This is also a good time to inquire about the composition of the department and whether you would be working with experienced engineers that you can use as mentors. Finally, if you don't have a clear idea about the type of work you would be doing if hired, ask them.

Send a follow-up e-mail immediately after the interview—Many interviewers wait a day or so before filling out their interview feedback form so their opinion may be positively influenced if they read your e-mail before filling it out. Use the e-mail to reiterate your interest in something you discussed in the interview or to correct something that you got wrong during the interview. If the letter does not specifically address the interview, for example, if it looks like a "one-size-fits-all" letter, it will be ignored.

10.1.4 Selecting the Right Offer

Hopefully your efforts will result in a list of job offers. Consider these points when selecting an offer.

Does the company offer job security?—In a tight economy, job security is an important consideration. It is a frequently held belief that a job with a larger company will provide job security, but this is not necessarily true. Large companies generally have multiple people who can do the same job, so it is easier for them to reduce their workforce, but their size can also make it easier for them to find a spot for you. On the other hand, smaller companies often rely on key individuals and retain them even during difficult times, but they are more subject to economic ups and downs. There is no right answer; you need to consider what you want and what is right for you.

The best way to stay continuously employed is to maintain and develop your skill set and experience base. In other words, job security is associated with you—not any company. If a company can provide you opportunities to gain experience and expand your technical knowledge, then you will develop job security. If, for some reason, this job doesn't work out, another company will value what you have learned; that is the best job security.[2]

Did the company offer a high starting salary?—Throughout your career you obviously want to be well compensated with a continuous pattern of salary increases. Your starting salary represents what the company believes is necessary to attract qualified new graduates. It is minimally, if at all, related to your unique talents. On the other hand, your salary increases and promotions will be based on your individual performance.[3] If we consider salary as a linear function of time, starting salary is the y-intercept and experience and technical knowledge represent the slope. To get the highest overall compensation, focus on gaining experience and expanding your knowledge rather than on your initial salary.

Did you meet mentors that you could learn from?—During the interview you might have asked about the composition of the department or about possible mentors, or you may have seen experienced engineers working in the laboratory. Mentors are an excellent way to quickly learn about the products and technologies at the company as well as the procedures for getting things done. A good mentor will also help you develop your overall engineering skills. Working with a mentor will accelerate your career.

Did you see women in engineering and management positions?—Modern workplaces are benefiting by hiring more women and promoting them at

[2]Throughout your career your skill set and experience will change, but it will always be your best source of security.

[3]Companies clearly understand that they must pay competitively to retain good engineers and they pay close attention to local salary surveys.

all levels. As a woman you will benefit from female mentors as well as an environment that values your contribution.

Is the laboratory well equipped?—Laboratory and troubleshooting skills are some of the first ones you want to pick up in your career. During the interview you'll likely get a brief tour of the facility. Pay close attention to the laboratory. If it is well equipped and there are engineers working in it, chances are you'll be able to work there also.

10.2 KEEPING A JOB

10.2.1 The First Year

Your first job is the beginning of a career that will ideally bring you many years of personal and professional fulfillment. During your first year it is important to establish your value to the company, establish good working relationships, increase your skill set, and begin to sort out your long-term career goals.

As a recent graduate you have a special opportunity. The author is biased, but typically the members of engineering departments are simply good people to work with. We may view you as a toddler who makes simple and laughable mistakes but is also learning at an incredible rate. As experienced engineers, we'll enjoy watching you make mistakes and learning from them; and we'll go out of our way to help you. Your responsibility is to accept help gracefully, work hard, and not ask anyone to do your work for you. As you mature, extend this gift to the recent graduates you encounter.

Learn to use the whiteboard in meetings—Every conference room has one or more whiteboards with dry-erase markers and erasers. You will likely spend at least an hour of every day conferring with other engineers in meetings. The most effective communicators are good at drawing diagrams and sorting out issues by drawing on the whiteboard. This is an acquired skill. At first, your drawings will be the wrong size, they'll be cramped, and after working with a drawing you'll wish you could erase it and start over. If you expect to use a diagram in a meeting, prepare a sketch first and use it as a guide, or go to the conference room a few minutes before the meeting and draw the skeleton of your drawing. It's always worthwhile to have a brand new dry-erase marker in your pocket when you go into a meeting because the ones in the conference room always seem to run out of ink just as you start to draw.

Learn to be a good troubleshooter—Troubleshooting is an invaluable skill at any company. If you are a good troubleshooter you'll be able to get your designs working quickly, and you'll be in demand when your group is working on system problems. Find a mentor who is a good troubleshooter, and don't be surprised if this person does not have an engineering degree.

As your career progresses, use your troubleshooting skills to become a good overall problem solver.

Document your work—Keep a paper or online lab notebook and make sure that all of your ideas, computations, circuits, meeting notes, and so on are kept in it. Before throwing away a piece of scratch paper check to see if it should be copied or taped into your notebook. Reference the relevant pages of your lab notebook in memos you write or in the comments of your code. Management hates to see engineers spend time retracing their steps, and appreciates when you can recall things quickly from your notebook. They especially appreciate it when a coworker digs into one of your notebooks because you referenced it in a memo or in a piece of code. Treat your notebook as a public document and limit its contents to technical and project issues. If one of your colleagues has a particularly good notebook, ask him/her to share his/her skills with you.

Learn more than what is required for anything you're assigned—You gain experience with every task you do, but you can greatly expand your experience base and probably do a better job if you make an effort to understand the bigger picture of what you're doing. By taking a small amount of extra time, you still get the job done, you remember the result better and you will broaden your skill set. This must be balanced against your workload. If you do it on your own time it will be an excellent investment in your career.

Always be productive—In a corporate environment, you'll have to accommodate the schedules of others. For example, you may not be able to proceed with a project until after a meeting or until someone else takes some data. One of the worst things you can tell your boss is that you are "waiting." Instead, maintain a prioritized list of short-, medium-, and long-term tasks so you can remain productive. This will make every hour of every day meaningful.[4] It will also establish your reputation as a self-directed worker.[5]

Include a test strategy with test criteria as part of any design—When you design something you should have a clear understanding of how you will test it and how you will know it meets the design requirements. This should be discussed at any design review so you know your expectations are aligned with the other team members. Frequently this process identifies new design requirements that can be included on the first prototype instead of being recognized at a later stage. It will also result in faster debug and a smoother transition into production.

Participate in IEEE activities—If you did not join the Institute of Electrical and Electronic Engineers (IEEE) in college, this is a great time to join and get involved. This organization is dedicated to helping you during every phase

[4]Sometimes the best 15-minute task is a cup of coffee with a colleague.

[5]If you enjoy this reputation, your supervisor will likely give you the latitude to direct yourself.

of your career. As a recent graduate you'll benefit by participating in the Graduates of the Last Decade (GOLD) program which will connect you with other young professionals in your area and allow you to expand your network of contacts. The IEEE also provides technical seminars where you can expand your technical horizons. Their flagship magazine "Spectrum" will keep you up-to-date on a superbly chosen collection of issues relevant to electrical engineers.

Don't nod your head if you don't understand—Recent graduates are often reluctant to tell a coworker or supervisor that they don't understand what they are being asked to do. If you nod, you are acknowledging that you understand what is expected and can complete the task. If you then fail, your supervisor will assume you weren't ready to accept the task. If you don't understand, let him/her know and develop a plan to come up-to-speed before receiving additional direction. A good way to do this is to get the information you need from a colleague and then return to your supervisor for additional details.

Develop writing and presentation skills—Before sending an e-mail, put yourself in the position of the reader. Gauge his/her technical level, how busy he/she is, and whether he/she even cares about what you're telling him/her. Then make sure your e-mail contains just the right amount of detail. Say you are sending an e-mail to your supervisor suggesting a meeting about an issue. You don't need to fully describe the issue; just give a clear reason why the meeting is needed. Internally evaluate the effectiveness of every e-mail you receive and every presentation you attend. Learn from others' mistakes and successes.

Make connections in other parts of the company—Limiting your interactions to your department will limit your understanding of how the company works and what it needs to be successful. Take advantage of any opportunity to interact with people from other departments and develop good working relationships with them. Sometimes companies create "cross-functional" teams to address problems that affect multiple departments. These are excellent opportunities to increase your value to the company by learning how other parts of the company work. Make sure that people from other departments are comfortable approaching you.

Ask your coworkers for help—As a recent graduate, you will find that most of your coworkers will be glad to share their knowledge and experience with you. However, busy engineers don't like being asked for help when the person asking has not taken the time to understand a problem or propose one or more alternatives. If you make an attempt to work out issues before asking a coworker you'll earn their respect—even if your approach is not quite right. If you graciously accept their assistance you'll learn faster and solve problems correctly the first time. If your boss trusts that you have the judgment to ask for help when necessary, he/she will trust you with more responsibility.

10.2.2 After the First Year

During your first year, you will gain the basic skills required to be a valuable member of your team. During the next few years you should take on additional responsibility, solidify your professional goals, and take steps to realize them.

Take on the worst job and make it the best job—There are some jobs that are important to the company but appear uninteresting, unglamorous, and, as a result, no one wants to do them. These projects may initially seem boring, difficult, or thankless. Some may be in a state of disarray because the previous engineer had either done them poorly or left the company. These jobs are often excellent opportunities to learn new skills, exercise creativity, and gain visibility in the company.

Recently, the author was asked to be the technical representative on a cross-functional team that reviewed the process that the company used to specify and procure toroidal inductors. The company was having problems with numerous suppliers and wisely realized that the underlying problem might be internal. The team consisted of representatives from engineering, purchasing, quality control, and manufacturing. As the engineering representative, the author helped develop an iterative process between the designer and the supplier during the design phase of a toroid. Then the author's team created design guidelines and worksheets to help engineers create designs that are manufacturable. Since their procedure was developed with input from multiple departments and toroid vendors, it was adopted as standard practice.

Working on this project had multiple benefits. The cost savings from this project gave their team an opportunity to present their results to the executive management, where they were favorably recognized. The project gave the author the opportunity to learn how things are done in other parts of the company and establish working relationships with the people who do them. It also allowed the author to show their quality control personnel some useful statistical techniques—and for them to introduce the author to the processes that they use.

Present new ideas to management and colleagues—If you believe you have a good idea, prepare a brief presentation describing your idea and discuss it with your colleagues or supervisor. Will it save time, money, or make the product more reliable? If it's not a good idea you'll learn why. Good managers appreciate enthusiastic ideas, even if they're off base.

Understand your relationship with each of your colleagues and optimize it—Shape your relationship with each colleague so you are both successful. Ask yourself these questions about each colleague: Do you respect this person? Do you consider him/her a mentor? Do you compete with him/her for projects? Would you consider him/her a friend? Do your skills and interests complement each other? Your colleagues may not realize that you are doing this, but they will regard you as someone who helps them succeed.

Learn to effectively delegate—You will encounter tasks in parts of your projects that are not right for you to do. These are not the tasks that you simply don't want to do. They are the tasks that can be done more efficiently by others, while your talents are better used elsewhere. For example, the author frequently designs hardware, but asks the company's computer-aided design (CAD) experts to draw schematics. This frees the author up to create a test strategy while the schematic is being drawn. As a result, the author is able to present the highest quality schematics as well as a test plan at the design review. Other engineers insist on drawing their own schematics and delegate the test strategy to a test engineer with equally beneficial results. Determine the delegation strategy that works for you.

If you're wrong, admit it and make it right—If you consistently check your work before passing it to others, you will enjoy a reputation for credibility. As a credible person, your colleagues and subordinates know that when you ask them to do something they will have to do it only once. If you constantly change your mind, or change directions, they will have to discard their efforts and then repeat them. If you get a reputation for wasting your colleague's time, they'll avoid working with you. When the author makes a mistake, I always apologize to those affected. If anyone has to work additional hours due to the my error, I make sure the person and their supervisor know that I was the one at fault.

If your career is technical get your PE license—Very few electrical engineers outside of the utility industry or government have their PE license.[6] If you passed the FE exam, you can take the PE exam after working for a year. This license distinguishes you from other engineers by showing that you have demonstrated competence in numerous areas of electrical engineering. If you intend to be a consultant in the future you should get your license because it allows you to testify in court and to use the legally restricted title of Consulting Engineer. The author found that the review manual and exam were similar to the approach used in this book; they emphasized practical skills but related them to the fundamentals. If you enjoyed the technical chapters of this book, you'll enjoy studying for the exam also. And when you pass it you'll be able to distinguish yourself with the initials PE on your business card!

Find the people who will be your long-term friends—Most experienced engineers have a collection of 5 or 10 current and former colleagues that they consider trusted friends and confidants. These are the people with whom

[6]Some electrical engineers believe that since their colleagues are not PEs, they will not be able to use them as references and their application will be rejected by the licensing board. The board recognizes this dilemma and will usually allow non-PEs to serve as your references.

you will share technical knowledge, solicit help from for difficult career or personal decisions, and might possibly form your own company one day. These friends are rare, and it is important that you identify them and then maintain the friendships as your careers progress.

Don't get overworked—Some engineers always seem to be overworked. They will often leave a job because they're "burned-out" and then find themselves burned-out at their next job. These engineers will insist that the company is overworking them, but that's rarely the case—especially when it happens to the same person at multiple companies. If you find yourself working an excessive number of hours, take a hard look at the reason why. It is easy to stay later and later at night, but the result of this can be an ever-increasing cycle of increasing hours and decreasing efficiency. It is much healthier to work efficiently for a reasonable number of hours and then recharge yourself.

Get to know the people who sell your company's product—The company's marketing and sales people have the knowledge and skills to convince customers to spend money on the products you make. Without sales, there is no company, so these people can easily be considered the most valuable in the organization. Good marketing people interact frequently with engineering. They may notice at a trade show that your competitor's product has a new feature that yours doesn't. A 10-minute conversation over a cup of coffee will let them know whether the feature is a simple enhancement or a complete product redesign. Alternatively, from your engineering vantage point you may come up with an idea for a new feature or product. If you present it to a marketing person he/she will be able to quickly tell you if he/she thinks it will sell. Finally, if you find yourself starting your own company in the future, you will need the best marketing people to present your product to paying customers.

Steer your career toward job satisfaction—You may have to change jobs a few times to end up in a position that is right for you. One of the author's friends, Andy, loved the ocean so, after college, he took a job at Scripps Institute of Oceanography in San Diego. Now he owns his own oceanographic company and combines his engineering skills with his love for the ocean. His company is small, but larger companies avoid competing with him because they simply can't match his passion. If you love football, find out who makes the system that draws the yellow first-down line on the televised image. Research how it works and tell them you want to work with this technology because you find it fascinating. If you like cars, learn about engine control or regenerative brakes and then send letters to car manufacturers. Don't waste good passion. Harness it in your career.

Don't take the promotion if you don't want to do the job—If you do your job well you will invariably be approached: "You're really good at what you do. We'd like to leverage your knowledge by having you manage a group that performs your function." For many engineers the promotion and corresponding salary increase is an excellent opportunity to move along their career path. But some engineers accept the promotion for the wrong reasons such as pressure from the management, for the salary increase, or because "that's what you're supposed to do." If you accept the position for the wrong reasons, you will be dissatisfied with your work, and you'll be a poor manager. If you believe your technical work is more fun than focusing on people and projects, save yourself, potential subordinates, and management the headache and politely decline the offer.[7]

10.3 ENJOYING YOUR WORK

You will be working for many years. If you enjoy your work, you will be successful at it and vice versa. The author hopes that the material in this book will merit valuable space on your bookshelf. You are encouraged to review the information in this section every few years. What is not meaningful now may be in a few years.

Take care of yourself—Personal fitness involves your mind, body, and spirit. Maintaining these will keep you at your peak of efficiency and creativity. During the writing of this book, the author rode his bicycle up a 1600 ft. peak several times per week, and after late evenings of writing, took long walks with the dog to wind down. Working hard and laughing with my colleagues gave me the energy to write for hours after long days at work. Find out what is required to take care of your mind, body, and spirit so you can perform at your best.

Your subordinates are as important as your superiors—If your superiors are wise then they value the opinion of your subordinates, and one of your duties as a senior person is to help your subordinates succeed. When you are leading the team, make sure the younger engineers get the credit they deserve. If you get a reputation for treating your subordinates well, you'll have no shortage of good people wanting to be on your team. As you mature further you may find that helping your subordinates provides more satisfaction than impressing your superiors.

Keep seeking out mentors—As a younger engineer you probably found mentors that you truly respected. As you gain experience, mentors are harder to find and you may note that your younger colleagues have skills and

[7]Companies often provide technical career paths in order to retain good technical personnel.

knowledge that you don't. As you mentor them in some areas, let them mentor you in others.

Don't get complacent or overly comfortable—If you look back over your career you'll likely find that you did your best work in stressful situations such as when you were struggling against a deadline or when a problem seemed insurmountable. If you are complacent or comfortable you are not working to your potential and could be a layoff target. Consider teaching a younger colleague to do tasks that are easy for you and then apply your experience to something more challenging. Does this make you feel uncomfortable? That's the point!

Laugh lots—There is no better time to laugh than when the team is stressfully attempting to do the impossible. A good laugh releases endorphins, makes us breathe more deeply, and reinforces the bond between teammates. Humor is a valuable part of the workplace, but must be used carefully. The author's favorite target of humor is himself which minimizes the possibility of offending anyone and often causes others to laugh at themselves also. When the group laughs together things just don't seem so difficult.

Listen to your colleagues—At this stage of your career you will be the one listening to and evaluating the ideas of your younger colleagues. By listening to them and carefully guiding their creativity, they'll value your inputs and ask you to participate in their efforts. Always encourage their enthusiasm.

Never stop learning—If you have maintained your technical skills, consider buying a textbook describing the newest technology in your field once a year and learning it. Despite the fact that textbooks contain hundreds of pages, chances are that most of those pages are review and you can focus only on what is new to you. It is not enough to skim the material and just learn the "buzzwords." Make sure you learn the material, work the examples and problems, and gain valuable skills that you can apply at work. Alternatively, take advantage of the many continuing education resources provided by the IEEE.

Avoid office politics by being competent—Fortunately engineers are usually great people to work with, and our enjoyment of our work, combined with a collective disdain for office politics makes personal skirmishes somewhat rare. Furthermore, as engineers, our output is objectively graded on our adherence to schedules and the quality of our work, and there is little room for subjective interpretation. The author's experience is that competence and usefulness to the company always seem to trump political scheming.

Share everything you know—Don't ever hoard knowledge in the hopes that it will make you uniquely valuable to the company. If you do this, your colleagues and management will recognize it immediately, obtain the knowledge via other means, and you will likely be terminated in the next layoff. Instead, use your experience and knowledge to learn new skills that are valuable to the company and to your subordinates. Be confident in your ability to learn and share all you know with your colleagues. This will make you both professionally valuable and well respected.

Value and maintain the friendships that you've made throughout your career—It is harder to be an older engineer than a younger engineer. At this point, you and your best friends will likely work at different companies. Keep in touch with these friends as you keep in touch with your non-engineer friends, share whatever new skills you're learning, and help each other sort out difficult issues. Take care of each other.

AFTERWORD

I had three goals in writing this book.

The first goal was to help you prepare for your immediate task of succeeding in interviews and getting a job.

The second goal was to set you on a path of maintaining and enhancing the skills that you have worked hard to acquire. Your skill set will change as your career progresses, but will always be the cornerstone of your value to any company.

The third goal was to share with you my own fascination, appreciation, and enjoyment of the field of electrical engineering. Thirty-five years ago, as a high-school senior, I narrowly selected this field over being a professional musician or a motorcycle mechanic. Though I still enjoy playing music and working on mechanical things, I have found great passion in my work and enjoy it immensely. Every day produces new design challenges; every year produces technology shifts that require new and interesting ways to apply the fundamentals; and every job position means finding new and special friends.

I hope the material in this book will merit a valued spot on your bookshelf. The technical material in the first nine chapters will help you tackle design tasks during your first years in the workplace; the material in the final chapter will help you apply your craft throughout your career in an effective and enjoyable manner. May it serve you as a valuable reference for many years!

Ten Essential Skills for Electrical Engineers, First Edition. Barry L. Dorr.

ANSWERS TO PROBLEMS

CHAPTER 1

1.1 $R_1 = 2941\ \Omega$, $R_2 = 1515\ \Omega$

1.2 $R_1 = 930\ \Omega$, $R_2 = 1400\ \Omega$

1.3 $R_1 = 85.24\ \text{k}\Omega$, $R_2 = 41.84\ \text{k}\Omega$, $R_3 = 23.43\ \text{k}\Omega$

1.4 For a current source, the current will remain constant regardless of the voltage so $Z = \mathrm{d}V/\mathrm{d}i = \infty$. The DC current source should be represented as an open circuit.

1.5 874.8 Ω

1.6 **(a)** Attenuation: V_O/V_{IN}, 1/3
(b) DC bias, 2.5 V
(c) Load impedance, 27 kΩ
(d) Source impedance, 6 kΩ

1.7 115 mA

1.8 624 mW

CHAPTER 2

2.1 $P_D = 16$ W

2.2 $T_J = 196$°C

Ten Essential Skills for Electrical Engineers, First Edition. Barry L. Dorr.

2.4 The junction temperature would rise to 193°C which would compromise the MTBF of the transistor.

2.5 The thermal resistance must be less than 5.87 °C/W so about 130 FPM is required.

2.6 2 J/°C

CHAPTER 3

3.1 Transfer functions for a through d are

$$H(s) = \frac{1}{RC} \cdot \frac{1}{s + \frac{1}{RC}}$$

$$H(s) = \frac{s}{s + \frac{1}{RC}}$$

$$H(s) = \frac{1}{RC} \cdot \frac{s}{s^2 + \frac{1}{RC}s + \frac{1}{LC}}$$

$$H(s) = \frac{s^2 + \frac{1}{LC}}{s^2 + \frac{R}{L}s + \frac{1}{LC}}$$

3.2 The transfer function is

$$H(s) = \frac{1}{R_2C} \cdot \frac{s}{s^2 + 2\xi\omega_n s + \omega_n^2}$$

where

$$\omega_n = \sqrt{\frac{1 + \frac{R_1}{R_2}}{LC}}$$

and

$$\xi = \frac{1}{2}\left(\frac{1}{R_2C} + \frac{R_1}{L}\right)\sqrt{\frac{LC}{1 + \frac{R_1}{R_2}}}$$

The frequency response is

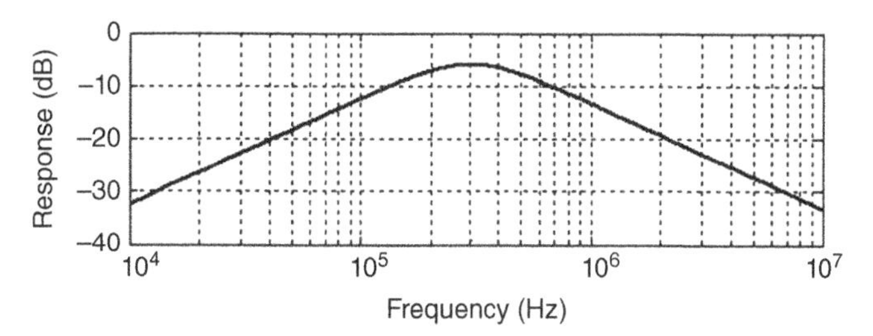

Frequency response plot for Problem 3.2.

3.3 The updated network is

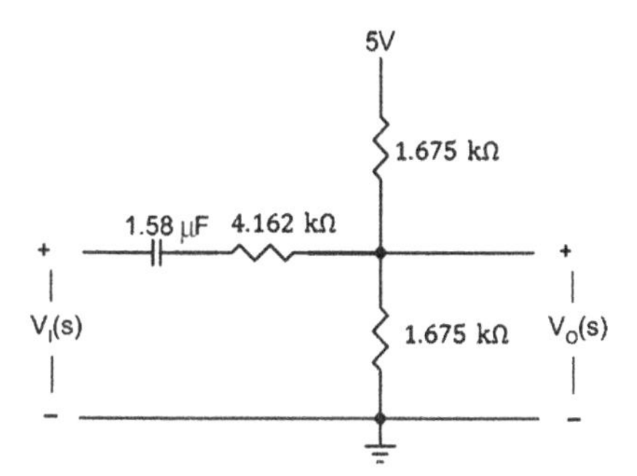

Updated network for Problem 3.3.

3.4 The transfer function is

$$H(s) = \frac{g_m A_V}{R_C C_B C_C} \cdot \frac{s}{s + \frac{1}{R_A C_A}} \cdot \frac{1}{s + \frac{1}{R_B C_B}} \cdot \frac{1}{s + \frac{1}{R_C C_C}}$$

where the networks from left to right are A, B, and C.

The frequency response is

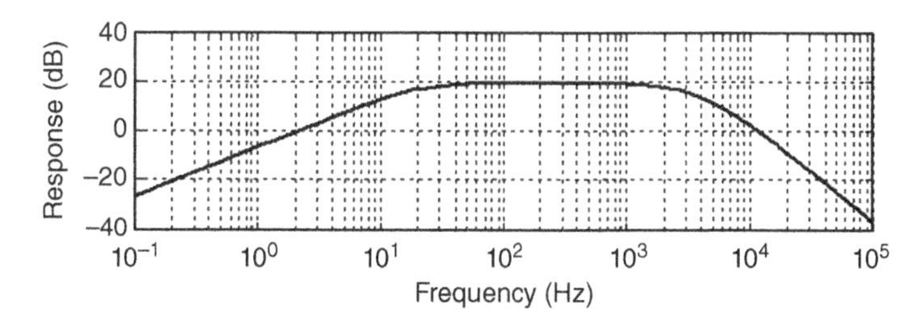

Frequency response plot for Problem 3.4.

3.6 See plot for Problem 3.2.

3.7 See plot for Problem 3.2.

3.8 Asymptotic analysis shows that the impedance at DC is $101 + j0\ \Omega$ and the impedance at infinite frequency is $96.6 + j0\ \Omega$.

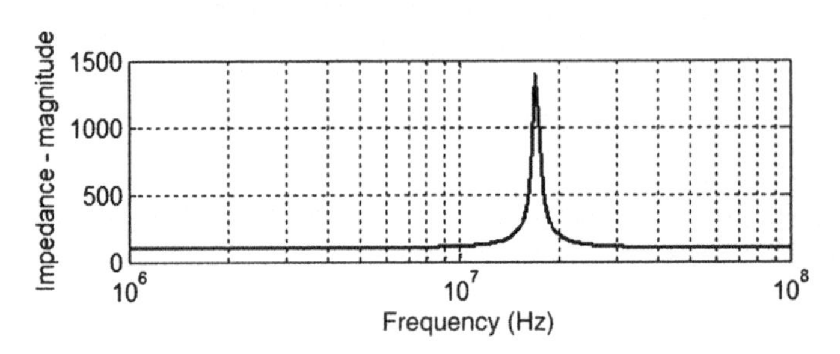

Input impedance versus frequency for Problem 3.8.

CHAPTER 4

4.1 $P = 1.03\%$

4.2 Probability of path stuck open is 1×10^{-8}. Probability of path stuck closed $= 2 \times 10^{-4}$.

4.3 $P = 9.38\%$

4.5 Three-sigma test limit is ± 212 mV. Tell the engineer that his test limits are excessively wide and would pass boards with incorrect or marginal components. Those boards could pass the test but then possibly fail in the field.

CHAPTER 5

5.1 $V_O(s) = V_{IN}(s)\dfrac{G(s)}{1 + G(s)H(s)} + D(s)\dfrac{1}{1 + G(s)H(s)}$

5.2 The setpoint is 0.71, the integrator output is 0.2. $G_C = 704 \times 10^{-3}$.

5.3 $\tau_2 = 96\ \mu s$, $T_S = 1.2$ ms P.O. = 25.4%

5.4 Double the gain by lowering the resistors in the integrator by half, that is, change them from 235 to 117.5 kΩ.

5.5 The sample rate is 79.6 kHz. The gain is 0.126.

CHAPTER 6

6.1 $V_O = \frac{R_2}{R_1}(V_B - V_A)$

6.2 $V_O = -\frac{R_2}{R_1}\left(V_{IN} - V_{Bias}\right) + V_{Bias}.$

If $V_{IN} = s(t) + V_{Bias}$ then the output is $V_O = -\frac{R_2}{R_1}s(t) + V_{Bias}$

6.3 The loop gain must be reduced by a factor of 10. The resulting gain for the noninverting amplifier is 10.

6.5 V_{IO} must be less than ±1.09 mV.

6.6 $\frac{V_O(s)}{V_{IN}(s)} = 3.5\left(1 + 3.17 \times 10^{-6}s\right).$

The values of resistance that move the zero to 100 kHz are $R_1 = 60.2$ kΩ and $R_2 = 150.6$ kΩ. R_2 is the feedback resistor.

6.7 Magnitude = 5.022, phase = −29.02°

CHAPTER 7

7.1 $\tau = 47\ \mu s$; $F_{3dB} = 3386$ Hz. Time to reach 0.87 V = 95.9 μs.

7.2 $H(s) = \frac{s}{s + \frac{1}{\tau}}$

7.3 Yes, because there are 5.2 octaves between the two frequencies. 6 dB per octave gives 31.2 dB.

7.4 Yes, the 3 dB frequency is 41,270 Hz. The attenuation at 820 kHz is 25.97 dB.

7.5 The normalized frequency is 3.23. A fourth-order Butterworth low-pass filter will provide about 40 dB attenuation at 42 MHz.

7.6 About 49.5 dB.

7.7 $N = 2$ filter satisfies the requirement. $R = 28.13\ \text{k}\Omega$, $C_1 = 1$ nF, $C_2 =$ 500 pF. Note: Analog devices tool gives 28.13 kΩ, but resistors can be any value provided capacitors are properly scaled.

7.8 Filter is shown below. Inductor Q of 10 causes 6.15 dB attenuation at 21 MHz. Passband attenuation is increased and the sharpness of the frequency response is reduced by the series resistance.

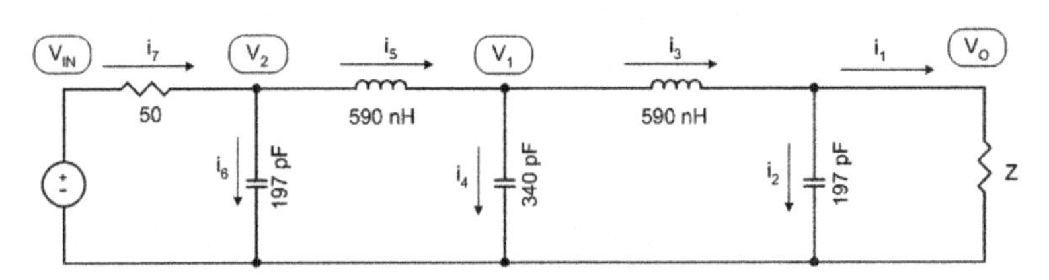

Filter for Problem 7.8.

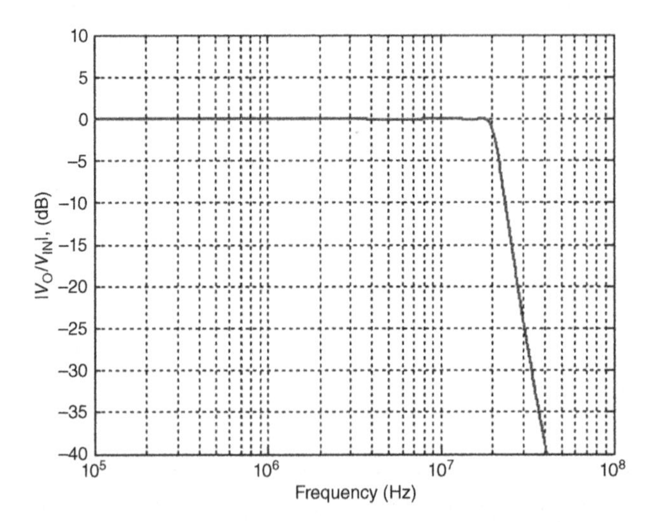

Frequency response plot for Problem 7.8 with Q = 1000.

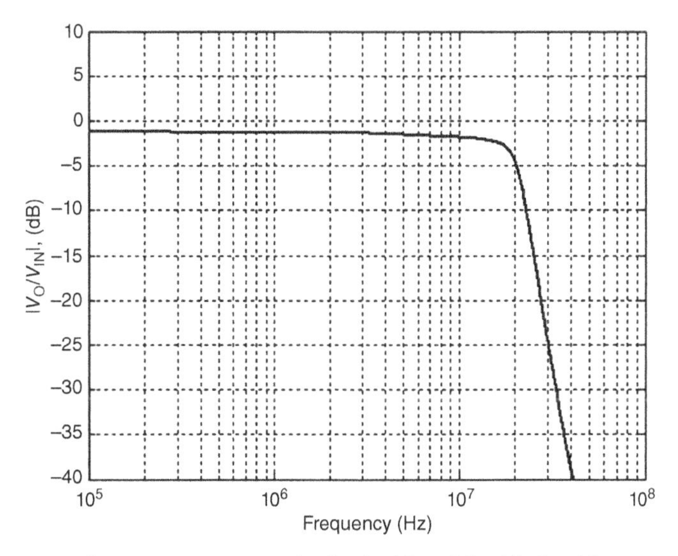

Frequency response plot for Problem 7.8 with Q = 10.

CHAPTER 8

8.1 **(a)** It would be passed.

(b) The frequency at the DAC output would be 1.5 kHz and would consist of steps because there is no reconstruction filter.

(c) The brick wall filter would cut off at 10.5 kHz and therefore reject the input signal. Nothing would be seen at the DAC output.

8.2 The filter rolls off at 12 dB per octave, so half the sample frequency must be three octaves above 100 kHz or 800 kHz. The signal must be sampled at 1.6 MHz.

8.3 The two frequencies are 100 kHz and 1 MHz. Use Equation 8.6. The signal at frequency f_2 is 0.574 dB smaller than the signal at frequency f_1.

8.4 **(a)** Transfer function is

$$H(z) = 0.5 \times \frac{\left(1 - a_2\right)\left(1 - z^{-2}\right)}{1 + a_2 z^{-2} + a_1 z^{-1}}$$

(b) DC gain is determined by substituting $z = 1$. It is zero.

(c) The frequency response is.

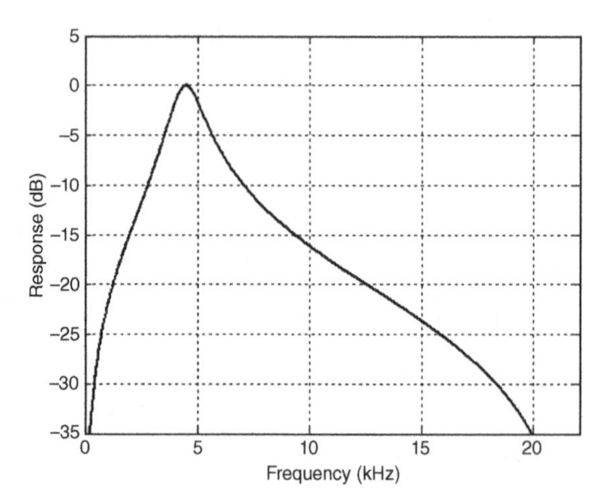

Frequency response plot for Problem 8.4.

8.5 Use $F_S = 480$ Hz with $N = 8$. Use Equation 8.37 to compute 26.23 dB at 63 Hz.

8.6 Use Equation 8.37 for the frequency response. Cascading the two filters provides more than 30 dB rejection in the stopband which is significantly better than a single filter:

$$|H(f)| = \frac{1}{5 \times 7} \left| \frac{\sin\left(\pi \times 5 \times \frac{f}{100{,}000}\right)}{\sin\left(\pi \frac{f}{100{,}000}\right)} \frac{\sin\left(\pi \times 7 \times \frac{f}{100{,}000}\right)}{\sin\left(\pi \frac{f}{100{,}000}\right)} \right|$$

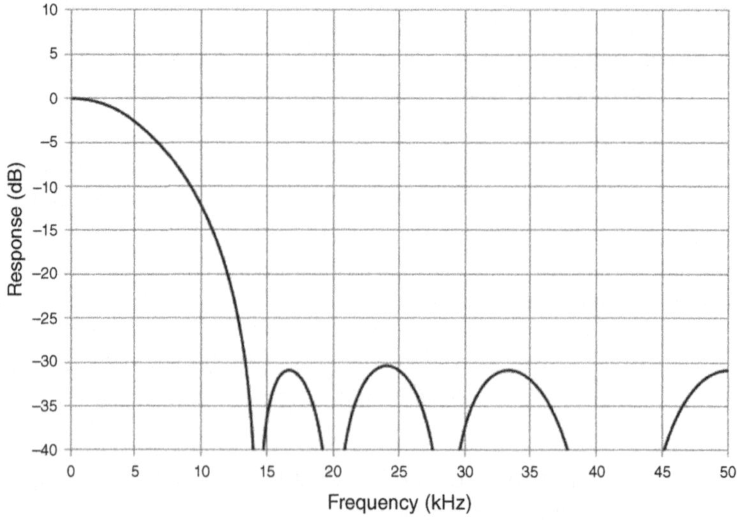

Frequency response plot for Problem 8.6.

8.7 Increasing k by 1 lowers the cutoff frequency by a factor of 2.

8.10 The transfer function is

$$H(z) = \frac{A(z)}{X(z)} = \frac{1}{1 - \alpha z^{-1}}$$

To get DC gain, substitute $z = 1$ which gives

$$\text{DC gain} = H(z) = \frac{1}{1-\alpha} = \frac{1}{1-0.9375} = 16$$

The wider register is used to accommodate the gain of 16. This is called "bit growth."

CHAPTER 9

9.1 **(a)** Reflections should be considered because the edge width is significantly less than the propagation delay.
(b) 2.56 V
(c) 0.838
(d) 4.71 V
(e) 2.15 V
(f) 2.96 V

9.2 **(a)** Yes, because at 125 ns the clock goes from 0.992 to 4.64 V
(b) At 875 ns the voltage at the load is 2.919 V. After that it remains higher than 2.9 V.
(c) Add 16.78 Ω source resistance.
(d) Add 551.12 Ω to ground at the load.

9.3 $Z_{IN} = 38.05 - j4.34\ \Omega$, $K_L = -0.1329 - j0.0558$

9.4 A Smith chart is not needed because the electrical length is negligible.

9.5 $Z_{IN} = 43.4 - j36.4\ \Omega$

9.6 31 pF, 0.166 μH

INDEX

Ten Essential Skills for Electrical Engineers, First Edition. Barry L. Dorr.